WRITING
MANUFACTURING PROCEDURES

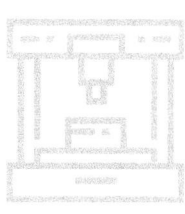

A CONTENT AND STRUCTURE GUIDE
JASON TESAR

TABLE OF CONTENTS

INTRODUCTION

WHY SHOULD YOU DOCUMENT YOUR PROCESSES?

Have you heard the saying, *you don't know how well you understand something until you try to teach it to someone else*? The discipline of documenting your processes can help identify information holes, instructional contradictions, process inconsistencies, and areas where your business is at risk. And afterward, the results of that effort (the documentation) will enable consistency in the training of personnel and the operation of your manufacturing processes, which will ultimately improve the consistency of your products or services.

This is the conceptual basis for Good Manufacturing Practice (GMP), Good Documentation Practice (GDocP), and quality certifications like ISO 9001.

Successful manufacturing companies:

1. **Say what they do** (document their processes)
2. **Do what they say** (hold themselves accountable to their documentation)

IS THIS THE RIGHT BOOK FOR YOU?

Perhaps your company has already documented its processes, but are you happy with the results? Are you a quality assurance manager who struggles with organizing the manufacturing section of your document control system? Maybe you're a production manager who needs to overhaul your department's documentation and retrain employees in order to reduce the frequency of mistakes. Perhaps you're a process engineer who understands the intricacies of how to build a product but not how to communicate that knowledge to others. Maybe you're a technician who wants to clean up existing documents because you have some proficiency at writing. Or are you an operator who wishes that your manufacturing instructions were written in plain English so they'd actually be usable?

Whatever your role in the manufacturing environment, you know that accurate documentation is *helpful* for learning and performing any repeated operation. And as the complexity of your company's product increases, usable documentation becomes not just helpful but *necessary*. If your company must comply with industry standards, *necessary* becomes *critical*. Perhaps even *legal*, depending on the industry in which you operate.

But documentation can also turn into a liability. Without a well-defined plan and consistent execution, instructions become duplicated. When one set of instructions is updated but the other forgotten, duplications begin to deviate from one another and eventually become contradictions. When new information is added, it ends up in the wrong document or a counterintuitive location that ensures it will quickly be forgotten. The number of challenges begins to add up when factoring in how many content creators and editors are involved. Over time, instructions that were intended to inform are ignored, and employees begin relying on training, word-of-mouth, or memory to perform their tasks correctly. Yet, without accurate documentation, the content of training varies from person to person. Word-of-mouth can't be trusted. And memory is notoriously unreliable. The frequency of mistakes increases. Someone should address the problem, but everyone is too busy building product (a money-making activity) to deal with documentation (which feels like a money-wasting activity). And even if the resources were actually made available to correct the problem, where would one start?

Hopefully your company's situation isn't as dire as what I've just described. But if you're a manager, engineer, technician, or operator working in a manufacturing environment and responsible for creating or updating documentation, this book can help you solve or avoid the most common problems.

WHY LISTEN TO ME?

In a manufacturing environment, the personnel who perform the work (engineers, technicians, and operators) don't necessarily have the organization or written communication skills to properly explain the complexities of what they do. And those tasked with documenting the work (often QA personnel or outsourced writers) may not have sufficient hands-on experience to describe the work itself.

I'm an expert technical writer and editor with over 20 years of experience in manufacturing. The majority of that time has been spent in an engineering-intensive industry, where controlled documentation systems are fundamental and adherence to government and international standards is required. I'm in the unique position of having performed the work and written about it, which means I thoroughly understand the challenges involved. I've also seen the positive effects of clear, organized communication.

In a well-designed documentation system, there is a place for everything and everything is in its place. Users understand the information and where to find it because the information is understandable and in a logical place. Such a system doesn't become a tangled mess, because when new requirements are added, content creators know exactly where to put the information. A well-designed documentation system is not a high-maintenance, resource-draining monster. It's a foundational tool for businesses to achieve consistency and quality. And these characteristics can help define a company's identity, making it stand out in the marketplace.

WHY DID I WRITE THIS BOOK?

Budgets are tight … and growing tighter every day. A manufacturing company may need a documentation system but lack the initial resources to put one in place. Not to mention the ongoing resources required to maintain it. This means the task of designing and creating a system will likely be handled internally, with minimal expense. If so, the structure of the system must be understood in advance. And documents must be written quickly but be thorough enough to hold their value for years to come.

So how can this be accomplished?

I wrote this book to answer that question by walking you through concepts and methods I developed while working in a variety of manufacturing environments (both low-volume/high-mix and high-volume/low-mix production, as well as manufacturing services). These methods are directly applicable to small and medium-sized businesses using a printed or simple electronic implementation, including those with controlled document systems. Even large businesses will benefit from the concepts in this book, particularly if those companies suffer from common, growth-related system problems (duplication of info and related discrepancies, inefficiencies associated with inconsistent format, poor system organization due to non-scalable system design, etc.).

I want to pass along my experience in order to make the design, creation, and maintenance of manufacturing documents as efficient and inexpensive as possible. Why? Because lack of resources needn't be a barrier to improvement.

WHAT CAN YOU EXPECT TO LEARN FROM THIS BOOK?

This book is not about the writing process. There are plenty of other books on the market that address the topics of interviewing stakeholders, outlining your proposed content, writing your first draft, seeking editorial input, revising, getting approval, and publishing your documents. This book is about the content itself—what information you'll need for a robust manufacturing documentation system and how you should structure that content to maximize usability and minimize redundancy.

The overall presentation in this book is from general principles to specific information. This is the case with the entire book, as well as within each section of the book. I'll explain the context for the information and then proceed to give you the details.

In the *GETTING STARTED* section, I'll discuss some high-level concepts you'll want to consider while designing a document system or getting to know your existing one. Before you begin creating documents, you'll need to understand how those documents will relate to each other and what functions they must perform. You'll also want to understand the types of information being conveyed through those documents and the audiences who will receive that information.

This will lead you to consider the structure of your company, and when you do that, you'll find that the information you intend to document has an inherent organization, which will become the pattern for your document system. Following this pattern, I'll introduce the three document types that will form the basis—and perhaps the entirety—of your manufacturing documentation.

In the *EQUIPMENT OPERATIONS PROCEDURE, INSPECTION CRITERIA SPECIFICATION*, and *PROCESS PROCEDURE* sections, we'll dive into the details of the three document types in the same order that you should create them. For each, I'll cover the following:

- What the document is
- Why it's needed
- Which employees will use it
- Where it fits in the document system
- How many are needed
- And most importantly … what content it should contain

Then I'll walk you through each section, giving you a detailed explanation of the content needed there. I'll present the information in the order in which I recommend arranging it within the document. If your company already has a controlled document system that dictates the topics and organization of your manufacturing documentation, the topics I present can be split up and arranged to comply with it.

Once you understand the three foundational document types, we'll look at some modifications, exceptions, and other documents you may need in the *OTHER DOCUMENT TYPES* section.

Everything up to this point will have involved structure and content. In *STYLE, FORMATTING, AND PRESENTATION*, we'll discuss how to present your content in the clearest manner possible.

When you reach the *NEXT STEPS* section, you'll have everything you need to get started creating your own documents. But if you want to speed up the process, I'll show you where you can download my document templates, which are structured and preformatted based on the principles in this book.

Finally, for those of you who are using Microsoft Word, I'll give you specific instructions on how to accomplish some of the most common formatting tasks in *FORMATTING WITH MICROSOFT WORD*.

GETTING STARTED

At their most basic level, documents are just information. So whether you're contributing to an existing document system or building one from scratch, it's helpful to understand some general concepts regarding informational structure. Although this book focuses on manufacturing, these concepts may also be applied to other departments.

TOP-DOWN PERSPECTIVE

Before writing any documentation, you need to know the purpose of your content, the scope of topics it will cover, and the intended audience. Will your content be instructive, merely informational, or both? Does this information already exist elsewhere? If so, how much of your content will be new information? Does the new information replace or supplement existing information? Does it contradict?

All of these questions arise because your documentation won't exist in a vacuum. It will have a relationship with other information. This larger informational structure is a system, and to know how your new documentation fits into it, you must first design the system or get to know the one that is already in place. This is what I call a top-down perspective.

What types of information already exist or need to be created? Are there different audiences for these information types? Are there areas of overlap between the audiences and information types? How can you organize the information to avoid duplicating it?

Understand the system conceptually, from the top-down. This is the same approach I've taken with this book—walking you through concepts and design considerations first, before discussing how to implement those concepts.

BOTTOM-UP PERSPECTIVE

After you understand the structure, you can begin building a new system, modifying an existing one, or just contributing content in an efficient and effective manner. When you reach this stage of the process, you'll be creating the individual pieces of a system and/or fitting those pieces together. And since you can't fit pieces together that don't exist in the first place, you must build or modify the system from the bottom-up. This is what I call a bottom-up perspective.

To reiterate ... design or learn the system from the **top-down**; build or modify it from the **bottom-up**.

EXTERNAL VS. INTERNAL FACING CONTENT

Press releases. Websites. Ads. All effective marketing content is created with an external-facing purpose. Its goal is to appeal to an external audience—potential customers. And its topics, organization, and language are structured to support this goal. To answer customer questions, meet their needs, and build their confidence in your business.

If you create an internal manufacturing document system with an external marketing mindset, the result will be a confusing mess.

Manufacturing documentation is different because the audience is internal to the company. Its topics, organization, and language must support a different goal. Because that goal is the proper functioning of the company, manufacturing documentation is most effective when it is patterned after the organization itself. In this way, the pieces of the document system are tangible versions of the company's functions.

THE FUNCTIONAL STRUCTURE OF A COMPANY

Within the company are **departments**, operating sequentially or in parallel, each performing a vital business function:

- The *Marketing* department attracts potential customers.
- The *Sales* department turns leads into paying work.
- The *Engineering* department develops the processes.
- The *Manufacturing* department builds the product.
- The *Maintenance* department sustains the equipment.
- The *Quality Assurance* department guarantees the function and condition of the product.

- The *Shipping and Receiving* department accepts raw materials and sends out finished product.
- The *Facilities* department maintains the environment in which the company operates.

Within these departments are **processes**, also operating sequentially or in parallel with each other and performing vital functions. These processes may be confined to a particular area of a manufacturing facility, distributed throughout, or may not have physical boundaries at all (such as an engineering team whose members work remotely).

In manufacturing, each process receives some type of input (raw materials or a partially built product), performs a function (adding, modifying, or removing material), and produces an output (an altered form of the product). Because of this, processes typically have storage areas for raw materials, processing stations and associated equipment, inspection stations and associated equipment, staging areas for incoming and outgoing product, and a multitude of other physical entities. Unless you have a fully automated assembly line (in which case your company has probably advanced beyond the point where this book is needed), these physical entities don't run themselves. They require people to operate them. And people require information.

INFORMATION LEVELS BY AUDIENCE

But what kind of information? That depends on the employee's role. The following are three roles typically found in a manufacturing environment, and as you will see, thinking through the intended audience should have a dramatic effect on the information you're attempting to convey.

> NOTE: The following roles and responsibilities will vary greatly between companies. These are presented only for the sake of discussion, so any document-related conclusions drawn from these should be tailored to how your company is actually organized.

- **Operators** perform the day-to-day functions of the process. They operate the equipment, perform inspections, and move product to the next station. They need to know how to operate the equipment, not how to program the machine itself or perform maintenance on it. They need to know whether a defect makes the product rejectable, not the minute details of how that defect will affect the product's functionality. They need to know how many products must be produced by the end of their shift, not how fluctuations in product flow will affect downstream processes.

- **Technicians** do the hands-on work of setting up and maintaining processes. They need to know how to operate the equipment as well as how to program it. They must understand why a given defect makes the product rejectable so they can give guidance on whether or not the inspection criterion applies. They need to understand the impact of one process on another so they can respond appropriately to variations in the quantity of products being produced.

- **Engineers** design and analyze the experiments that lead to the processes in the first place. They don't need to understand the details of how to operate the equipment, but they need to understand its capabilities and how those capabilities compare to similar machines. They must know the minute details of how a product functions so they can ensure that all the necessary attributes are being inspected. And they often interface with customers about production quotas, so they must also understand the impact of product losses at various points along the manufacturing line.

If operators must sift through information intended for engineers to find what they need, the likelihood of error increases and the manufacturing line loses efficiency. Personnel should have all the info they need, when and where they need it, and nothing more. To accomplish this objective, the information should be specific to the roles and responsibilities present within your company.

THE NATURAL SEQUENCE

When you consider the information you intend to convey, as well as your audience, it becomes clear that patterning your documentation after the organization itself is the best choice. But with so much information to consider, where should you begin your documentation efforts?

As we discussed in *The Functional Structure of a Company*, there is an inherent pattern to the organization of a company. In the same way, if you were to draw a diagram to represent the flow of product from one process to another on your manufacturing line, you'd quickly notice another pattern—a sequence.

Granted, it may be complex, with parallel paths, decision loops, and such. But at a high level, there would be a beginning, middle, and end as depicted in *Figure 1*.

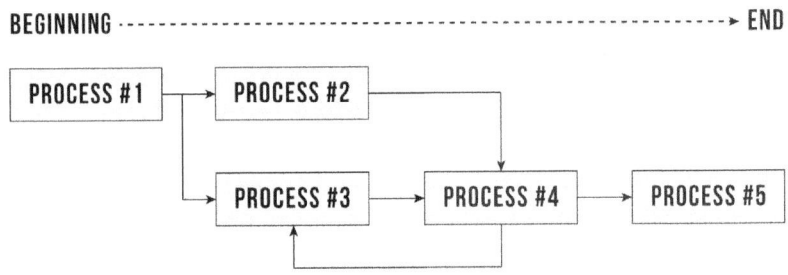

Figure 1: The Manufacturing Line

As you can see, the arrangement of processes in a manufacturing line also has an inherent structure to it. Each process is its own entity, yet it has a relationship to other processes. Whether or not the manufacturing facility is physically arranged to support this structure, the documentation can be. And doing so should be among your first considerations.

Now, if you were to take any one of these processes and summarize the basic activities involved in running it, they would be:

1. Understanding what, when, where, how, and why to do something
2. Doing it
3. Evaluating if it was done correctly

These roughly correlate to the first three steps of the *Plan, Do, Check, Act* (PDCA) cycle used for continuous improvement. Whether or not you are familiar with this methodology, the important point to note here is that running a process requires three separate but related tasks.

Employees running a manufacturing process must understand the details of what they are supposed to do. Then they do it, often with the aid of some equipment. Then they—or someone else—evaluates what they've done to ensure it was done correctly. These three tasks are the basic building blocks of manufacturing documentation.

PROCESS, EQUIPMENT, AND INSPECTION

If you were to break down each of your manufacturing processes into the categories of process instructions, equipment instructions, and inspection criteria, you would begin to understand how many times, or in how many locations, a given piece of information is used.

The more often a piece of information is duplicated, the greater the likelihood of those duplications becoming different from each other. If the information must be updated, will you know where to find all those duplications? If an employee finds what they need in one document, but a more up-to-date version resides in another document, the product might be built incorrectly. This results in rework, which increases costs and decreases the efficiency of the manufacturing line (by decreasing output).

In database design, one of the fundamental principles is to break down information into its smallest meaningful form. For instance, rather than storing a person's full name in a data table cell, best practice dictates that you would store a person's first and last name in different cells. This allows you to sort the information in the table by either first or last name, and it allows you to use each piece of information separately, if needed.

Manufacturing documentation doesn't need to be broken down to this level of detail, but the principle of *smallest meaningful form* still applies. If a certain make/model of equipment is used in ten different processes, do you need ten different instruction manuals for how to operate it? No; you need one manual that is available to the personnel in those ten locations. Should you have the inspection criteria for all attributes of every product in one giant specification? No; the criteria should be broken down and distributed to the personnel who perform those inspections.

Segmenting your information into the categories of **process**, **equipment**, and **inspection** allows you to share information across multiple processes instead of duplicating information that could later lead to contradictions and even scrapped product. This three-document structure is comprehensive enough to capture all the information needed to run the process, without excess or redundant information. And it's flexible enough to work for any combination of scenarios:

- One process using one machine
- One process using many machines
- One machine used by many processes
- Many processes using the same inspection criteria

EQUIPMENT OPERATIONS PROCEDURE

Now that we understand the information system from a top-down perspective, the difference between external and internal content, the inherent structure of a company, the level of information appropriate for different audiences, the natural sequence of manufacturing information, and how to segment that information, it's time to begin building the system from a bottom-up perspective.

WHAT IS IT, AND WHY IS IT NEEDED?

As its name implies, an equipment operations procedure is a document containing instructions for operating a piece of manufacturing equipment. Because most manufacturing involves the use of at least one machine, this procedure is one of the three basic documents that will comprise a robust informational system.

Equipment manufacturers often provide thorough documentation with the purchase of a machine. A user manual, programming instructions, electrical and mechanical schematics, and maintenance and repair information are commonly included. With all of this information readily available from the equipment manufacturer, you may wonder why an operations procedure needs to be written at all. The answer lies in the usability of that information.

The equipment manufacturer's information is a great place to start when building a document system, because it contains instructions on a variety of critical topics. But the breadth and depth of the information may not align with the roles and responsibilities of the positions at your company. Left in its original form, it can be a distraction at best and dangerous at worst. For this reason, the information should be extracted, reorganized, and translated for appropriate audiences.

In addition, you'll need a mechanism for storing and communicating institutional knowledge about the equipment—the tips and tricks that only become apparent after operating it for some time. The equipment operations procedure serves this purpose.

WHO USES IT?

There are many responsibilities in a manufacturing environment, and each company organizes them differently when it comes to roles, or employee positions. For the sake of efficiency, I will assume that your company follows a structure similar to this:

- **Operator** – performs daily operations on the equipment
- **Technician** – understands daily operations, and performs basic programming
- **Engineer** – understands daily operations and basic programming, performs advanced programming, and is responsible for equipment capability and output
- **Maintenance Technician** – understands daily operations and basic programming, and performs repair and maintenance

Notice that each of these roles requires knowledge of the daily equipment operations. This is why an operations procedure is a critical piece of the manufacturing documentation system—all of these employees use it. Also notice that an operator doesn't necessarily need to understand how to program or repair the equipment. Such topics should be covered in other documents.

For the equipment operations procedure, extract only the information needed to turn on the machine, set it up, run it, and shut it down. Then translate this information into common sense instructions for operators.

WHERE DOES IT FIT IN THE DOCUMENT SYSTEM?

The equipment operations procedure is one of the foundational pieces of a manufacturing documentation system, which is why I've chosen it as the starting point for discussion. In a hierarchal document system, it would be situated at the bottom as shown in *Figure 2*.

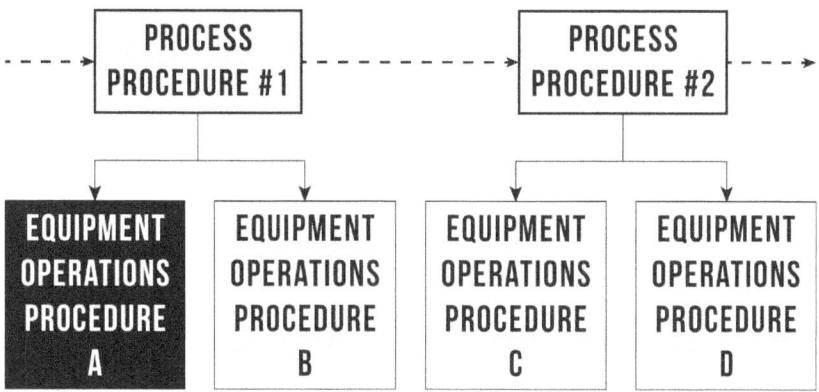

Figure 2: The Equipment Operations Procedure

HOW MANY ARE NEEDED?

Only one operations procedure is required per make and model of equipment. A different model of the same machine type—even if it performs the same function—may operate differently, in which case it would need its own procedure. Likewise, a machine from a different manufacturer will require its own procedure. However, two or more machines from the same manufacturer, with the same model number, can be operated using the same procedure.

The way that a machine functions doesn't vary. It has certain features or components, and those components operate the same way, regardless of who is utilizing them. What vary are the end results of using the machine. One department may utilize components A and B to accomplish some outcome, while another department utilizes components C and D to accomplish something else. This type of information is comprehended in the *PROCESS PROCEDURE* and would be inappropriate for an equipment procedure.

If the equipment operations procedure is thorough enough to cover all the functional components of a machine, writing one document per make and model will prevent you from duplicating information and allow for the possibility of that same procedure to be used by multiple processes and/or departments.

HOW SHOULD IT BE STRUCTURED?

Given the possibility that the machine may be used by more than one process or department, it makes sense to structure the document around the functional components of the machine. These modules of instruction can then be arranged in sequential order, if there is a particular sequence in which the modules must be performed. If not, or if the sequence varies depending on what's being accomplished by the machine, arrange the modules in whatever order makes the most sense (alphabetical by function, in order of usage frequency, etc.)

WHAT CONTENT SHOULD IT CONTAIN?

Now that we've covered the introductory questions about this type of document, it's time to walk through the sections of content that will be required. If your company has a controlled document system, you will likely have to work within some constraints regarding section numbers and what types of information go in which sections. Or maybe you have free rein to organize the content in whatever way you see fit. Regardless, the following sections will cover all the content needed for a robust equipment operations procedure.

DOCUMENT IDENTIFICATION

The first thing a reader will want to know is whether or not they're reading the correct document. This question can be answered with a title page, headers and footers, or some combination of these. Whichever format you choose to deliver the answer, this is the information that should be presented:

- **Title** – a concise name that matches the subject matter of the document. Consider using the following format, replacing the bracketed content with your own:

  Procedure for [Equipment Manufacturer] [Make and Model] Operations

- **Document Number** – a numeric (or alphanumeric) designator that sets the document apart from any other, including others with similar titles

- **Revision** – a letter (or date code) that designates the revision of the document (to ensure that the correct version is being used)
- **Author**, **Reviser**, and/or **Owner** – the name of the person who authored the document, the name of the last person to update the document, and/or the name of the person or department responsible for its content (used for internal or external auditing purposes)

TABLE OF CONTENTS

The next thing a reader will want to know is where they can find the specific information they need within this document. A table of contents answers that question by giving an overview of the content and structure of the document.

It can also function as an abbreviated or (mini) procedure if the document headings are written as action statements (discussed in later sections) and styled appropriately throughout the document.

NOTE: You can read more about *Numbered Text and Levels* styling on page *83*.

After the content of the document is written and organized into a logical structure (also discussed in later sections), create a table of contents showing all the document topics and the high-level actions to accomplish them.

NOTE: You can read more about using a *Table of Contents* on page *82*.

PURPOSE, SCOPE, AND AUDIENCE

A title may not be sufficient to convey the breadth of information contained in a document. This section, which follows the table of contents, allows for elaboration about the document's purpose, the scope of its content, and its intended audience.

Purpose

Oftentimes, a reader will go directly to this section to determine if the document contains what they need. State its purpose in clear language, so readers know whether or not they should keep reading.

Consider using this wording, replacing the bracketed content with your own:

> "The purpose of this procedure is to describe the operating instructions for the [equipment manufacturer] [make and model] used for [purpose of the equipment]."

Scope

After a document system is designed, its structure is well understood by those who designed it. The same can't always be said of those who actually use the system on a daily basis. They may not care about the structure, so long as they can find the information they need to do their jobs. For the sake of reinforcing the system's design, it's beneficial to define the scope of the document so readers know what type of content it does and does not cover.

Consider using this wording:

> "The scope of this procedure is limited to equipment operating instructions. It does not include process instructions, inspection criteria, or requirements specific to customers, products, or product revisions."

Audience

Defining the audience of the document helps readers know if the content applies to them. Beyond that, it also serves as a guide for the author—a target to keep in mind while creating the content. Knowing the audience is vital to creating useable documentation.

Consider using this wording:

> "The audience of this procedure is personnel trained to operate, program, or maintain this equipment."

DOCUMENTS AND FORMS

In manufacturing environments, documents typically function in relation to other documents in some capacity. Other documents may act as prerequisites to the one you're writing—your content won't make sense unless the reader first understands some other document's content. Or perhaps your new document points readers to forms that they're required to fill out. Whether or not you choose to separate these into subcategories (*Required Documents*, *Referenced Documents*, *Forms*, etc.) or combine them into a single section, it is helpful to provide a summary of all the documents referenced in the one you're writing. Users can then see at a glance what other documents are involved, and the task of auditing your document system will be that much easier.

1. Create a bullet list of all the referenced documents and forms, beginning with their document numbers, followed by their titles.

    a) Don't include the revision numbers, as doing so would only require the references to be updated each time the revisions change. Numbers and titles don't change often, so the list should remain accurate with little maintenance.

2. Arrange the list in ascending document number order, so readers can quickly look up the information they need.

3. Consider using the following format, replacing the bracketed content with your own:

    • [*Document Number*], [*Document Title*]

Another benefit of having a document list is so that each entry can be turned into a bookmark. Instead of manually typing the reference—and struggling to remember the document number and its exact title and capitalization—you can just insert a reference to it. Each instance of that cross-reference will then be consistent in spelling, wording, punctuation, and capitalization. Also, inserting a cross-reference will carry over the styling (notice the italicized style in the suggested format above, which helps to set the cross-reference apart from normal text).

> NOTE: You can read more about *Setting up Bookmarks* on page *95* and *Inserting Cross-References* on page *95*.

After the content of your document is written, come back to this section and run a text search on each document number. If there is only one instance of a document reference (the one in this section), you'll know you forgot to mention some critical information. Or perhaps you'll reconsider whether that document really needs to be on the list. Delete it if it doesn't.

> NOTE: You can read more about searching for *Text Strings* on page *98*.

DEFINITIONS

Each industry has its own acronyms, abbreviations, jargon, and slang. Within that industry, each company may also have its own version of those terms. Use this section to summarize all the applicable terms that show up in the document, but don't fall into the trap of trying to create a comprehensive dictionary. Only include the information that is relevant to the document.

1. Create a bullet list of all the terms.
2. Organize the terms alphabetically.
3. After each term, write a definition.
4. Format the term (not the definition) in some manner that will set it apart from regular text. Consider using the following format, replacing the bracketed content with your own:

    *   [*Term*]: [Definition]

5. Set up the term (whatever comes before the colon in the example above) as a bookmark.
6. Insert a cross-reference to this bookmark anywhere this term is used throughout the document.

After the content of your procedure is written, come back to this section and run a text search on each term. If there is only one instance of a term reference (the one in this section), you'll know you forgot to mention some critical information. Or perhaps you'll reconsider whether that term really needs to be on the list. Delete it if it doesn't.

SAFETY PRECAUTIONS

For many reasons (administration, auditing, liability, etc.), it is beneficial to have all safety-related information in one place. However, precautions are most effectively implemented when listed throughout the document at the locations where the topic is relevant. For these reasons, the safety precautions section should be structured similarly to the previous *Documents and Forms* and *Definitions* sections, where the summary list and all the individual warnings throughout the document are synchronized to each other. The only difference is that the wording of the summary entries cannot be made into bookmarks, because they will vary slightly from the corresponding warnings, due to the lack of context from the surrounding instructions.

For example, in the *Equipment Setup* section, you might warn the reader about a pinching hazard when installing a particular module on the equipment. In the summary section, that warning won't make sense unless you add wording at the beginning of the entry to recreate that context, such as:

"During setup, when installing the …"

Mechanical, Electrical, Chemical, Ergonomic

Use this section to create a bullet list of all the mechanical, electrical, chemical, ergonomic, and other precautions specific to operating the equipment.

Do not include general precautions related to the surrounding environment, unless those precautions are not stated in a general workplace safety document. If those are already stated in a general document, repeating them here duplicates the information, enabling one set of warnings to become obsolete or to contradict the other. It also creates the possibility of an employee following one and not the other, which could lead to injuries due to lack of critical information.

After the document content is written, running a text search on phrases within each summary entry (ones that correspond word-for-word with their warnings) will help ensure that no safety topic is left out.

> NOTE: You can read more about searching for *Text Strings* on page *98.*

If all warning statements are formatted correctly, running a text search on each instance of the word *warning* will also help ensure that the summary list is complete.

> NOTE: You can read more about formatting *Notes, Cautions, and Warnings* on page *84.*

Personal Protective Equipment (PPE)

Use this section to create a bullet list of all the personal protective equipment referenced in this procedure, the part number (your company's or the manufacturer's, if applicable), as well as a definition of what each item is used for.

1. Format the items in some manner that will set them apart from regular text. Consider using the following format, replacing the bracketed content with your own:

    *   [*Item name*] [(Part number)]: [Purpose of the item]

2. Set up the item name (whatever comes before the parentheses in the example above) as a bookmark.
3. Insert a cross-reference to this bookmark anywhere this item is mentioned throughout the document.

    > NOTE: You can read more about *Setting up Bookmarks* on page *95* and *Inserting Cross-References* on page *95.*

After the content of your document is written, come back to this section and run a text search on each item. If there is only one instance of an item referenced (the one in this section), you'll know you forgot to mention some critical information. Or perhaps you'll reconsider whether that item really needs to be on the list. Delete it if it doesn't.

SPECIAL HANDLING

Special handling involves precautions where the equipment or product is at risk of damage. Similar to the *Safety Precautions* section, the same reasons exist for having all special handling information summarized in one place as well as being listed throughout the document at the locations where the topic is relevant. The summary and the individual cautions should be synchronized to each other. Again, the wording of the summary entries cannot be made into bookmarks, because they will vary slightly from the corresponding cautions, due to the lack of context from the surrounding instructions. Make sure to recreate that context by adding wording at the beginning of each entry, such as:

"When loading product onto the conveyor belt …"

Mechanical, Electrical, Chemical

Use this section to create a bullet list of all the mechanical, electrical, chemical, and other special handling requirements specific to operating the equipment. Concentrate on instances where the equipment or product could be damaged by mishandling.

Do not include risks to the product unless they are specific to the equipment. Risks that are more closely related to certain products (or their revisions) would be more appropriate to define in customer-specific documents.

> NOTE: You can read more about *Customer-Specific Documents* on page *75*.

After the document content is written, running a text search on phrases within each summary entry (ones that correspond word-for-word with their cautions) will help ensure that no handling topic is left out.

If all caution statements are formatted correctly, running a text search on each instance of the word *caution* will also help ensure that the summary list is complete.

> NOTE: You can read more about formatting *Notes, Cautions, and Warnings* on page *84*.

EQUIPMENT

Primary and Backup Equipment

Use this section to create a bullet list of the equipment to which this procedure applies, and any backup equipment, if applicable.

Consider using the following format, replacing the bracketed content with your own:

- [Manufacturer] [make and model] [function of the machine or equipment]

Equipment is not the same as tools (which are defined in the following *Materials and Tools* section). Think of equipment as installed—or non-portable—machinery, including air hoses, fluid dispensers, and any other items installed at or near the workstation where the equipment is located.

However, there may be fixtures specific to this machine, or optional components of the machine that can be removed and installed as needed. If so, define them in whichever section makes the most sense, given the structure of your assembly line.

Equipment Overview

During the setup, operation, and shutdown of this equipment, users will interact with its various features, such as access panels, physical switches, control buttons, digital displays, user interfaces, and software menus. For the sake of clarity, most of these items should be visually introduced at specific sections throughout this procedure where the subject matter becomes relevant, but it's beneficial to first show the reader what the equipment looks like and to identify the general features that may be referenced over and over again.

Use this section to present images of the equipment and its general features, annotated to clearly define the names of these features so there is no confusion about what is being referenced. At a minimum, identify the location of Emergency Off (EMO) switches referenced in the *Safety Precautions* section.

> NOTE: You can read more about using *Figures* on page *84*.

MATERIALS AND TOOLS

Materials

Use this section to list of all the consumable items used during operation of this equipment, such as process gasses (CDA, N_2, O_2), solvents (IPA, Acetone, Lacquer Thinner, Mineral Spirits), and materials used for cleanup (wipes, cotton swabs, etc.).

These materials should not be confused with any that may be added to (or installed on) the product using this machine (screws, wires, epoxy, etc.). Such materials are part of the product's bill of materials (BOM) and should be identified in the *PROCESS PROCEDURE* or *Customer-Specific Documents*.

Create a bullet list of all the materials referenced in this procedure, the material's part number (your company's or the manufacturer's, if applicable), and a definition of what each item is used for.

1. Format the materials in some manner that will set them apart from regular text. Consider using the following format, replacing the bracketed content with your own:

   - [*Material name*] [(Part number)]: Purpose of the material

2. Set up the material name (whatever comes before the parentheses in the example above) as a bookmark.
3. Insert a cross-reference to this bookmark anywhere this material is mentioned throughout the document.

   > NOTE: You can read more about *Setting up Bookmarks* on page *95* and *Inserting Cross-References* on page *95*.

After the content of your document is written, come back to this section and run a text search on each material. If there is only one instance of a material referenced (the one in this section), you'll know you forgot to mention some critical information. Or perhaps you'll reconsider whether that material really needs to be on the list. Delete it if it doesn't.

| NOTE: | You can read more about searching for *Text Strings* on page *98*. |

Tools

As implied in the *Equipment* section, tools can be defined as portable—not installed—instruments used during operation of this equipment. These may include tweezers, rubber mallets, handling aides, and a multitude of other items, depending on your manufacturing environment.

These are not to be confused with equipment maintenance tools, unless operators are expected to perform basic maintenance operations and instructed to do so in this document. In such a case, consider creating a separate subsection for the maintenance tools.

Use this section to create a bullet list of all the tools referenced in this procedure, the tool's part number (your company's or the manufacturer's, if applicable), as well as a definition of what each tool is used for.

1. Format the tools in some manner that will set them apart from regular text. Consider using the following format, replacing the bracketed content with your own:

    *   [*Tool name*] [(Part number)]: [Purpose of the tool]

2. Set up the tool name (whatever comes before the parenthesis in the example above) as a bookmark.
3. Insert a cross-reference to this bookmark anywhere this tool is mentioned throughout the document.

    | NOTE: | You can read more about *Setting up Bookmarks* on page *95* and *Inserting Cross-References* on page *95*. |

After the content of your document is written, come back to this section and run a text search on each tool. If there is only one instance of a tool referenced (the one in this section), you'll know you forgot to mention some critical information. Or perhaps you'll reconsider whether that tool really needs to be on the list. Delete it if it doesn't.

EQUIPMENT SETUP

We come now to the first of several instructional sections (*Equipment Setup*, *Equipment Operation*, and *Equipment Shutdown and Cleaning*). Instead of lists and general information, these will present step-by-step directions for the reader to follow during setup, operation, and shutdown of the equipment. Because sequences are implied by these tasks, numbered lists will make up the majority of the content in these sections.

Outlining

Before writing any instructions, it's a good practice to plan your content using an outline. Start by brainstorming the high-level actions required to set up the equipment, such as:

- Turning on the power
- Starting the software
- Logging in to the software
- Loading the program
- Etc.

For each of these actions, consider what is required to complete it. For example, *loading the program* might involve:

- Checking your lot documentation for the product and revision identifiers
- Navigating to the appropriate directory and subdirectory on the equipment or company server
- Selecting the file name that matches the product and revision being run

Do this for the entire setup process. Make sure to specify when and how readers are to use other documents, such as filling out a form or logging setup data. Then organize all the information into topics, subtopics, and instructions.

Flowcharting

Presenting this information to a reader in a text-based format requires organizing it into a sequential flow. This type of presentation works well if the information itself is sequential, or linear, in structure. But what if it isn't? What if your instructions have decision loops and optional content, based on whether the product meets or fails to meet certain conditions? This is where the efficacy of a text-based presentation breaks down.

If the content cannot be easily represented by a standard outline of topics, subtopics, and indented instructions, you may need a flowchart to visually introduce your non-linear information.

If you decide to use a flowchart:

- Follow standard flowchart conventions regarding shapes and their meanings.
- Use the basic *Action Phrases* (discussed in the next subsection) as the content of your flowchart shapes, and remove all articles (a, an, the) to make the phrases as concise as possible.
- Label each flowchart shape with its corresponding section number, so the flowchart functions as a visual table of contents.
- Place the flowchart at the beginning of a section, as a way of visually introducing the content to the reader and warning them of its non-linear structure.
- The flowchart should flow left-to-right and top-to-bottom, like the text in a document.
- To ensure readability of the flowchart, limit the number of flowchart shapes across (left-to-right). If needed, the flowchart can be many rows long (top-to-bottom) to accommodate the length of the procedure.
- Present the flowchart as you would any other figure in your document.

> NOTE: You can read more about using *Figures* on page *84* and *Inserting a Table of Contents* on page *91*.

Action Phrases

Whether or not you use a flowchart, your outline of the *Equipment Setup* activities will need to be refined for clarity and precision. Starting with the high-level actions (topics), edit them as follows:

- Rewrite them as concise verb phrases, instead of topics or titles (i.e. "Turn on the equipment" instead of "Turning on the equipment" or "How to turn on the equipment"). These phrases will become the headings of your instruction list. If you used a flowchart at the beginning of the section, these phrases should match those in the flowchart shapes with the addition of articles (a, an, the).
- Number each phrase to convey the correct sequence of operations to the reader. If you used a flowchart at the beginning of the section, ensure that the section numbers match the flowchart shape numbers, including any decision points, conditional content, or iterations.
- Format the phrases as second-level headings so the table of contents (which displays all first and second-level content) will contain these abbreviated instructions, thereby functioning as a mini-procedure.

> NOTE: You can read more about using a *Table of Contents* on page *82* and *Numbered Text and Levels* on page *83*.

Step-by-Step Instructions

Indented beneath each action phrase, list the instructions for accomplishing that action, and edit them as follows:

- Rewrite them using active (instead of passive) language. For example, use "Rotate the power switch to the ON position," instead of "The power switch is rotated to the ON position." This may seem like common sense, but passive language often shows up in procedural documents because it is the default, conversational tone that authors use in their everyday life. In such a setting, passive language comes across as polite; in a procedure, it implies latitude where none may exist.
- Number the instructions to convey the correct sequence to the reader.
- Format the instructions as third-level text.

> NOTE: You can read more about *Numbered Text and Levels* on page *83*.

Other Considerations

In terms of the structure and content of the information, there are a few other considerations.

If the instructions to accomplish an action are used more than once in the sequence, don't repeat them. Define the instructions where they first apply, then insert a cross-reference to that action at the other locations in the sequence where it also applies.

NOTE: You can read more about *Inserting Cross-References* on page *95*.

If conditional instructions are so lengthy that readers will have forgotten the standard instructions by the time they get back to them, consider moving the content to an appendix at the end of the document and inserting a cross-reference to it. If the conditional instructions result from a problem (a deviation from the normal, expected sequence), move the content to the *Troubleshooting* section (discussed on page *37*), and define all the steps necessary to resolve the issue.

When pointing readers to a specific section of another document (such as inspecting a product feature per an *INSPECTION CRITERIA SPECIFICATION* or filling out a form field), use the exact wording of the section header, action phrase, criterion, or form field in that document. If such references are used often, consider creating a list of these references in the *Appendices and Miscellaneous* section (discussed on page *38*), setting them up as bookmarks, and inserting cross-references to them. Doing this maintains a one-to-one relationship between the instructions listed here and the wording used in other documents.

NOTE: You can read more about *Setting up Bookmarks* on page *95*.

EQUIPMENT OPERATION

This section of your equipment procedure should be constructed in the same manner as the *Equipment Setup* section, utilizing flowcharts, action phrases, and step-by-step instructions. Whereas the previous section defined how to prepare the equipment for operation, use this section to define how to operate the equipment or use it for its intended purpose.

Though similar in structure to the *Equipment Setup* section, this section may be larger and more complex, depending on the level of the machine's sophistication. A manually operated machine involves more actions on the part of the operator. And each of those actions must be defined, along with points of decision, the options available at each decision, the expected results, and where to find the appropriate instructions for when the results are unexpected (see *Troubleshooting* below).

On the other hand, for an automated piece of equipment, operating it may involve nothing more complicated than pressing the **Run** button.

EQUIPMENT SHUTDOWN AND CLEANING

The *Equipment Shutdown and Cleaning* section of this procedure should be constructed in the same manner as the *Equipment Setup* and *Equipment Operation* sections, utilizing flowcharts, action phrases, and step-by-step instructions. Use it to define how to shut down the equipment or return it to the state in which the reader found it prior to following the setup instructions. This would include any disassembly and cleaning activities that an operator would be expected to perform.

If the machine requires different shutdown methods based on when it will be used next, make sure to define these instructions under their own headings (short-term versus long-term shutdown, daily versus weekly, etc.).

TROUBLESHOOTING

Anyone who has operated a piece of manufacturing equipment knows that machines don't always do what is expected of them. This is because their automated actions are based on other variables, such as the condition of the materials being fed into them. Material handling systems will jam. Pattern recognition systems will fail to find fiducials. Software will encounter logic errors. Mechanical components will break. While the *Equipment Operation* section defines the smooth, intended sequence of events, the *Troubleshooting* section defines the exceptions.

Use this section to define the most commonly encountered equipment errors and the proper responses to them. Present the information by topic (following the same content guidelines used in the *Equipment Setup* section), with a summary table (such as an Out-Of-Control Action Plan used in the *PROCESS PROCEDURE*), or whatever method makes the most sense for your business. Regardless of which method you use, make sure to include the following:

- The symptom or problem
- Where in the operations sequence that problem might occur
- The possible causes of that problem
- The proper sequence of responses to each cause
- Who is responsible for performing the responses in that sequence

APPENDICES AND MISCELLANEOUS

The previous sections should cover most (if not all) of the information needed to operate a piece of equipment in a manufacturing environment. In the event that some topic—specific to your unique environment—was not covered, feel free to add it where appropriate. If it will not fit in the previous sections without interrupting the sequence and confusing the reader, consider adding it here at the end of the procedure.

Appendices are a great way to cover the following additional information:

- Content too lengthy to be included in the normal sequence of activities
- Conditional content
- Reference diagrams and/or photos
- Advanced programming, troubleshooting, or maintenance instructions (if not comprehended in a different document)
- Any topic not included previously, yet comprehensive enough to warrant its own section

INSPECTION CRITERIA SPECIFICATION

WHAT IS IT, AND WHY IS IT NEEDED?

As its name implies, an inspection criteria specification is a document that contains requirements used for evaluating product. Because most manufacturing involves some level of evaluating the quality of the product being produced, this specification is the second of the three basic documents that will comprise a robust informational system.

NOTE:	I use the term *inspection* in a generalized manner throughout this section. Although visual examination might be implied by this term, *inspection* also includes any evaluation performed during the manufacturing cycle (taking measurements, using a go/no-go gauge, etc.).

Fortunately, you may not have to create one from scratch. In industries such as automotive, electronics, and medical device manufacturing, standards have been developed by leading companies (sometimes in conjunction with the government) and made available for public use. These standards define a wide variety of criteria that acceptable product must meet, as well as the test methods to determine acceptability (in some cases).

With all of this information readily available through published standards, why does an inspection criteria specification need to be written at all? As we discussed with the equipment operations procedure, the answer again lies in the usability of that information.

Depending on your company's role within the industry, these standards may be viewed as merely guidelines or absolute requirements. Regardless of which, the breadth and depth of the information may not align with the types of product produced—or the inspection methods used—by your company. Left in their original form, the standards could be a distraction at best or dangerous at worst. For this reason, the information should be extracted, reorganized, and translated for appropriate audiences.

WHERE IS IT USED?

There are many possible inspection points in a manufacturing environment, and each company will implement those inspections differently in terms of which department is responsible and when the inspections take place. For the sake of discussion, I will assume that your company follows a structure similar to this:

- **Receiving Inspection** – evaluating the quality of incoming materials from vendors
- **Process Inspection** – evaluating the quality of product produced at certain processes; performed by manufacturing personnel, which could be considered a conflict of interest
- **QA Inspection** – evaluating the quality of products at various points along the assembly line; audit-style inspection, intended to eliminate conflicts of interest
- **Shipping Inspection** – final inspection of product prior to packing and shipment

Notice how often inspections occur, and that these roles don't necessarily take place within the same department. This is why an inspection criteria specification is another critical piece of the manufacturing documentation system—its use cuts across many employees and departments when implemented strategically.

HOW MANY ARE NEEDED?

A careless implementation of inspection criteria would be to refer each department and process to a single, comprehensive industry standard, without attempting to segment or translate the information. This makes each department and/or inspector:

- Responsible for finding all the criteria that apply to their inspection, regardless of where (and in how many locations) those criteria are located
- Responsible for properly interpreting and applying each and every criterion to their inspection

Such a scenario provides ample opportunity for criteria to be missed, overlooked, and/or misinterpreted, which could lead to defective product being shipped to a customer. For this reason, it's wise to extract the applicable criteria, segment it according to your business needs, and translate it into language appropriate for the persons performing the inspections.

When deciding how to segment the information, consider whether or not certain criteria will be used at multiple points along the manufacturing line. It's quite possible that some product feature evaluated prior to shipment, might also be evaluated at previous manufacturing processes ... perhaps even during receiving inspection if that feature is intrinsic to the incoming material. In such a scenario, in what ways would those inspections vary? Would different departments be involved? Would different pieces of equipment be used each time?

Taking into account these types of questions, the implementation that will generally allow for the most efficient use of criteria is to:

- Create one, standalone criteria specification for each product attribute (or group of attributes) that is **not** intrinsically linked to a particular process.

> NOTE: A process-independent criteria document is what I'm describing in this section.

- For the remaining criteria, embed them within the procedure of the particular process to which they **are** intrinsically linked.

> NOTE: You can read more about process-specific criteria in the *Process Inspection* section of the *PROCESS PROCEDURE* on page *67*.

HOW SHOULD IT BE STRUCTURED?

Industry standards are typically organized by informational topic, with the criteria grouped by whatever method seemed logical to the author(s). But in manufacturing, companies are not producing informational topics; they're running processes and producing product, which is how the information should also be organized.

If a product is inspected at only one point along the manufacturing line, the inspection is—by definition—intrinsic to that process. The criteria in that inspection don't apply to other processes (or departments), so there's no reason to separate them from the process procedure. Embed these process-specific criteria in the process procedure.

However, if a product is evaluated at more than one point along the manufacturing line, the criteria are—by definition—not intrinsic to any particular process. In such a case, the criteria may remain the same across processes (or perhaps departments), but the following may differ:

- Evaluation Method
- Evaluation Equipment
- Evaluation Frequency

For example, **100%** of the products in a manufacturing lot might be **visually inspected** with the **naked eye** at Process #1 for a particular attribute, **100% visually inspected** with a **microscope** at Process #5, and **measured** on a **5% sample** basis with **calipers** at Process #10.

In this example, because the variation occurs not in the criteria but the inspection itself, the methods, equipment, and frequency should be embedded in their respective process procedures. Only the criteria are consistent across processes, and should therefore be documented in a manner that would support their independence from any one process. These criteria should be separated by product attribute (or group of attributes) and defined in standalone specifications.

This may sound like an extreme situation, and may not currently apply to your business. But by structuring your documentation in this way, you can avoid duplication problems in the future as you add product lines, manufacturing processes, and departments.

After embedding all the process-specific criteria and inspection methods in their respective procedures, and creating standalone documents for all process-independent criteria, all the applicable industry standards should be represented in your documentation.

The next step would be to apply these same rules to any general criteria your company has internally developed, or industry standard criteria that your company has chosen to alter. Add these criteria into the appropriate process or product attribute document. If new standalone specifications are needed, create them.

The only other type of criteria remaining should be customer/product/revision-specific criteria, which is a separate matter that we'll discuss in *Customer-Specific Documents* on page *75*.

WHERE DOES IT FIT IN THE DOCUMENT SYSTEM?

In a hierarchal document system, the inspection criteria specification would be situated at the bottom as shown in *Figure 3*.

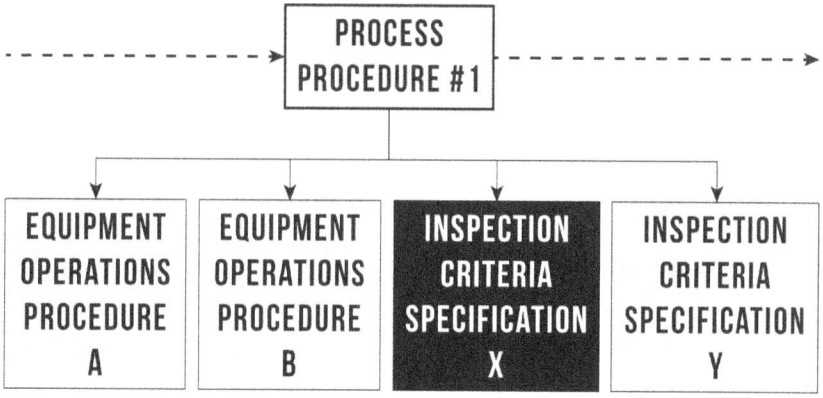

Figure 3: The Inspection Criteria Specification

WHAT CONTENT SHOULD IT CONTAIN?

Now that we've covered the introductory questions about this type of document, it's time to walk through the sections of content that will be required. If your company has a controlled document system, you will likely have to work within some constraints regarding section numbers and what types of information go in which sections. Or maybe you have free rein to organize the content in whatever way you see fit. Regardless, the following sections will cover all the content needed for a robust criteria specification.

DOCUMENT IDENTIFICATION

The first thing a reader will want to know is whether or not they're reading the correct document. This question can be answered with a title page, headers and footers, or some combination of these.

Whichever format you choose to deliver the answer, this is the information that should be presented:

- **Title** – a concise name that matches the subject matter of the document. Consider using the following format, replacing the bracketed content with your own:

  Inspection Criteria for [Product Attribute or Category]

- **Document Number** – a numeric (or alphanumeric) designator that sets the document apart from any other, including others with similar titles
- **Revision** – a letter (or date code) that designates the revision of the document (to ensure that the correct version is being used)
- **Author**, **Reviser**, and/or **Owner** – the name of the person who authored the document, the name of the last person to update the document, and/or the name of the person or department responsible for its content (used for internal or external auditing purposes)

TABLE OF CONTENTS

The next thing a reader will want to know is where they can find the specific information they need within this document. A table of contents answers that question by giving an overview of the content and structure of the document.

After the content of the document is written and organized into a logical structure (discussed in later sections), create a table of contents showing all the product attributes and criteria applicable to those attributes.

NOTE: You can read more about using a *Table of Contents* on page *82*.

PURPOSE, SCOPE, AND AUDIENCE

Purpose

Oftentimes, readers will go directly to this section to determine if the document contains what they need. State its purpose in clear language, so readers know whether or not they should keep reading.

Consider using this wording, replacing the bracketed content with your own:

> "The purpose of this specification is to define the inspection criteria for the [product attribute] portion of [applicable product types]."

Scope

After a document system is designed, its structure is well understood by those who designed it. The same can't always be said of those who actually use the system on a daily basis. They may not care about the structure, so long as they can find the information they need to do their jobs. For the sake of reinforcing the system's design, it's beneficial to define the scope of the document so readers know what type of content it does and does not cover.

Consider using this wording:

> "The scope of this specification is limited to inspection criteria. It does not include equipment operating instructions or process information."

Audience

Defining the audience of the document helps readers know if the content applies to them. Beyond that, it also serves as a guide for the author—a target to keep in mind while creating the content. Knowing the audience is vital to creating usable documentation.

Consider using this wording:

"The audience for this specification is personnel trained to inspect for these criteria."

APPLICABLE DOCUMENTS

As previously mentioned, manufacturing documents typically function in relation to other documents in some capacity. But if you attempted to list all the process procedures that reference this criteria specification (lateral references, given the hierarchal system examples in *Figure 2* and *Figure 3*), the list would be difficult to keep up-to-date. The only documents necessary to mention in this section are those industry standards to which this document complies.

Consider using this wording, replacing the bracketed content with your own:

"The criteria in this specification comply with the requirements in the following standards relating to [product attribute]:"

1. Create a bullet list of all the referenced documents, beginning with their document numbers, followed by their titles.

   a) Don't include the revision numbers, as doing so would only require the references to be updated each time the revisions change. Numbers and titles don't change often, so the list should remain accurate with little maintenance.

2. Arrange the list in alphabetical order, so readers can quickly look up the information they need.
3. Consider using the following format, replacing the bracketed content with your own:

   • *[Document Title]*, *[Document Number]*

Another benefit of having a document list is so that each entry can be turned into a bookmark. Instead of manually typing the reference—and struggling to remember the document number and its exact title and capitalization—you can just insert a reference to it. Each instance of that cross-reference will then be consistent in spelling, wording, punctuation, and capitalization. Also, inserting a cross-reference will carry over the styling (notice the italicized style in the suggested format, which helps to set the cross-reference apart from normal text).

NOTE: You can read more about *Setting up Bookmarks* on page *95* and *Inserting Cross-References* on page *95*.

After the content of your document is written, come back to this section and run a text search on each document number. If there is only one instance of a document reference (the one in this section), you'll know you forgot to mention some critical information. Or perhaps you'll reconsider whether that document really needs to be on the list. Delete it if it doesn't.

DEFINITIONS

Each industry has its own acronyms, abbreviations, jargon, and slang. Within that industry, each company may also have its own version of those terms. Use this section to summarize all the applicable terms that show up in this criteria specification, but don't fall into the trap of trying to create a comprehensive dictionary. Only include the information that is relevant to this document.

1. Create a bullet list of all the terms.
2. Organize the terms alphabetically.
3. After each term, write a definition.
4. Format the term (not the definition) in some manner that will set it apart from regular text. Consider using the following format, replacing the bracketed content with your own:

    - [*Term*]: [Definition]

5. Set up the term (whatever comes before the colon in the previous example) as a bookmark.
6. Insert a cross-reference to this bookmark anywhere this term is used throughout the document.

After the content of your document is written, come back to this section and run a text search on each term. If there is only one instance of a term reference (the one in this section), you'll know you forgot to mention some critical information. Or perhaps you'll reconsider whether that term really needs to be on the list. Delete it if it doesn't.

NOTE: You can read more about searching for *Text Strings* on page *98*.

GENERAL INFORMATION

Criteria Applicability

For the sake of the reader's understanding, it may be necessary to explain how the criteria in this document are arranged. Perhaps there are several related product attributes, with different criteria for each, while a few criteria apply to all the related attributes. Or maybe you wish to make a distinction between industry standard criteria and those developed internally. Use this section to define how and why the rest of the document is organized as it is, and which sections apply to which attributes.

Quality Level

As you will see in the next section, it is sometimes necessary to modify a criterion based on the product's quality level. A product that your company sells to a government contractor will likely require a more stringent inspection than one sold to a commercial retailer. These differences in quality levels must be recognizable to those performing the inspection operations. Use this section to define what those quality levels are and how to identify them.

Consider using the following wording:

> "Because reject criteria vary by quality levels, both the level and its associated criteria are listed together in tables throughout this document. The following are explanations of each quality level and its associated criteria:"

Follow this with subsections for each quality level, defining:

- The name of the quality level (prototype versus production, commercial versus military, etc.)
- How to identify manufacturing lots that use this level

- Which sections of the industry standard (or criteria) are comprehended in this level, if applicable

INSPECTION CRITERIA

Use this section to define all the criteria that apply to the product attribute in question. The information can be presented to the reader in any format (table, list, etc.), as long as the criteria and the needed accept / reject conditions are clearly defined. Whichever format you choose, this is the information that should be presented:

- **Defect Name** – a concise name that uniquely identifies the defect, preventing it from being confused with similar defects. Consider using the following format, replacing the bracketed content with your own:

  [Product attribute] – [Defect name]

- **Description** – additional information about the defect, or clarification of any terms used in the defect name
- **Quality Level** – the quality level to which the reject condition applies; used in conjunction with the reject condition (listed below)
- **Reject Condition** – the precise condition that, if met through observation or measurement, qualifies the product as rejectable; used in conjunction with the quality level (listed above)
- **Reference Photos / Diagrams** – photos or diagrams that help the reader visualize any of the above information

PROCESS PROCEDURE

WHAT IS IT, AND WHY IS IT NEEDED?

Now that the equipment operations procedure and inspection criteria specification are defined, it's time to connect these foundational pieces with a higher-level document that will explain how and when to use them. The process procedure is the third and final piece of the core document group that will comprise a robust informational system. As its name implies, it's a document containing instructions for running a process within a manufacturing environment, which includes many different topics, such as:

- How to verify that the previous process was completed
- How to prepare the workstation to start the current process
- Which raw materials are used in the process, where they are located, and in what quantities they are used
- Which machines are used to perform the process
- In what sequence product flows from one machine to the next
- At what locations, and in what quantities, products are staged between machines
- At what points inspections take place
- What data needs to be collected during inspections and what to do with that data
- What inspection methods (visual versus measurement) and frequencies (sample versus 100%) are used
- How to rework products that fail inspection
- How to respond and whom to contact when the process doesn't go as planned
- How to move the product to the next process

While the other two document types contain content that could be used by other processes or departments, the process procedure covers everything else that is intrinsic to the process itself.

WHO USES IT?

As with the other document types, the process procedure is used by several individuals:

- **Operator** – performs day-to-day functions of the process
- **Technician** – maintains the process, including making adjustments based on product inspections
- **Engineer** – analyzes the process and its data to determine opportunities for quality or throughput improvements

WHERE DOES IT FIT IN THE DOCUMENT SYSTEM?

While a process procedure might be one of the first concepts to discuss when designing a documentation system, it might be one of the last to be written (see *Top-Down Perspective* and *Bottom-Up Perspective* on page *12*). This is because it ties the other documents together and defines how and when to use them, which is only possible if those other documents already exist. In a hierarchal document system, the process procedure would form the top section as shown in *Figure 4*.

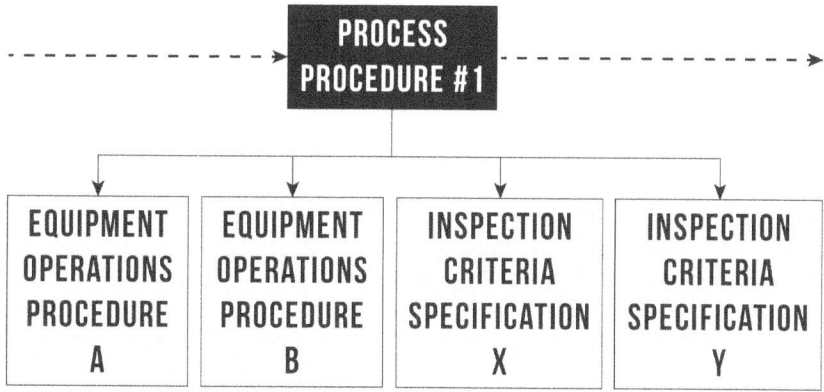

Figure 4: The Process Procedure

HOW MANY ARE NEEDED?

Only one procedure is required per process. Using more than one would confuse readers by breaking up the sequence of instructions.

HOW SHOULD IT BE STRUCTURED?

The primary goal of the process procedure is to communicate the correct sequence of operations from start to finish, within a given process. This is accomplished by presenting the information in chronological order and limiting exceptions and other distractions so the order remains clear in the mind of the reader.

This can be a difficult task when a process involves many decision points and conditional content, but the following sections will explain how to include all the relevant information without confusing the reader.

WHAT CONTENT SHOULD IT CONTAIN?

Now that we've covered the introductory questions about this type of document, it's time to walk through the sections of content that will be required. If your company has a controlled document system, you will likely have to work within some constraints regarding section numbers and what types of information go in which sections. Or maybe you have free rein to organize the content in whatever way you see fit. Regardless, the following sections will cover all the content needed for a thorough process procedure.

DOCUMENT IDENTIFICATION

The first thing a reader will want to know is whether or not they're reading the correct document. This question can be answered with a title page, headers and footers, or some combination of these. Whichever format you choose to deliver the answer, this is the information that should be presented:

- **Title** – a concise name that matches the subject matter of the document. Consider using the following format, replacing the bracketed content with your own:

  [Process name], Procedure for

- **Document Number** – a numeric (or alphanumeric) designator that sets the document apart from any other, including others with similar titles
- **Revision** – a letter (or date code) that designates the revision of the document (to ensure that the correct version is being used)

- **Author**, **Reviser**, and/or **Owner** – the name of the person who authored the document, the name of the last person to update the document, and/or the name of the person or department responsible for its content (used for internal or external auditing purposes)

TABLE OF CONTENTS

The next thing a reader will want to know is where they can find the specific information they need within this document. A table of contents answers that question by giving an overview of the content and structure of the document.

It can also function as an abbreviated or (mini) procedure if the document headings are written as action statements (discussed in later sections) and styled appropriately throughout the document.

> NOTE: You can read more about *Numbered Text and Levels* styling on page *83*.

After the content of the document is written and organized into a logical structure (also discussed in later sections), create a table of contents showing all the document topics and the high-level actions to accomplish them.

> NOTE: You can read more about using a *Table of Contents* on page *82*.

PURPOSE, SCOPE, AND AUDIENCE

A title may not be sufficient to convey the breadth of information contained in a document. This section allows for elaboration about the document's purpose, the scope of its content, and its intended audience.

Purpose

Oftentimes, a reader will go directly to this section to determine if the document contains what they need. State its purpose in clear language, so readers know whether or not they should keep reading.

Consider using this wording, replacing the bracketed content with your own:

> "The purpose of this procedure is to describe the process instructions for the [process name] process, used to [the purpose of the process]."

Scope

After a document system is designed, its structure is well understood by those who designed it. The same can't always be said of those who actually use the system on a daily basis. They may not care about the structure, so long as they can find the information they need to do their jobs. For the sake of reinforcing the system's design, it's beneficial to define the scope of the document so readers know what type of content it does and does not cover.

Consider using this wording:

> "The scope of this procedure is limited to process instructions. It does not include equipment operating instructions or requirements specific to customers, products, or product revisions."

Audience

Defining the audience of the document helps readers know if the content applies to them. Beyond that, it also serves as a guide for the author—a target to keep in mind while creating the content. Knowing the audience is vital to creating usable documentation.

Consider using this wording:

> "The audience of this procedure is personnel trained to perform, maintain, or support this process."

DOCUMENTS AND FORMS

In manufacturing environments, documents typically function in relation to other documents in some capacity. Other documents may act as prerequisites to the one you're writing—your content won't make sense unless the reader first understands some other document's content. Or perhaps your new document points readers to forms that they're required to fill out. Whether or not you choose to separate these into subcategories (*Required Documents*, *Referenced Documents*, *Forms*, etc.) or combine them into a single section, it is helpful to provide a summary of all the documents referenced in the one you're writing. Users can then see at a glance what other documents are involved, and the task of auditing your document system will be that much easier.

1. Create a bullet list of all the referenced documents and forms, beginning with their document numbers, followed by their titles.
   a) Don't include the revision numbers, as doing so would only require the references to be updated each time the revisions change. Numbers and titles don't change often, so the list should remain accurate with little maintenance.
2. Arrange the list in ascending document number order, so readers can quickly look up the information they need. Consider using the following format, replacing the bracketed content with your own:

   • [*Document number*], [*Document title*], *Procedure for*

Another benefit of having a document list is so that each entry can be turned into a bookmark. Instead of manually typing the reference—and struggling to remember the document number and its exact title and capitalization—you can just insert a reference to it. Each instance of that cross-reference will then be consistent in spelling, wording, punctuation, and capitalization. Also, inserting a cross-reference will carry over the styling (notice the italicized style in the suggested format above, which helps to set the cross-reference apart from normal text).

NOTE: You can read more about *Setting up Bookmarks* on page *95* and *Inserting Cross-References* on page *95*.

After the content of your document is written, come back to this section and run a text search on each document number. If there is only one instance of a document reference (the one in this section), you'll know you forgot to mention some critical information. Or perhaps you'll reconsider whether that document really needs to be on the list. Delete it if it doesn't.

> NOTE: You can read more about searching for *Text Strings* on page *98*.

DEFINITIONS

Each industry has its own acronyms, abbreviations, jargon, and slang. Within that industry, each company may also have its own version of those terms. Use this section to summarize all the applicable terms that show up in the document, but don't fall into the trap of trying to create a comprehensive dictionary. Only include the information that is relevant to the document.

1.   Create a bullet list of all the terms.
2.   Organize the terms alphabetically.
3.   After each term, write a definition.
4.   Format the term (not the definition) in some manner that will set it apart from regular text. Consider using the following format, replacing the bracketed content with your own:

   - [*Term*]: [Definition]

5.   Set up the term as a bookmark.
6.   Insert a cross-reference to this bookmark anywhere this term is used throughout the document.

After the content of your procedure is written, come back to this section and run a text search on each term. If there is only one instance of a term reference (the one in this section), you'll know you forgot to mention some critical information. Or perhaps you'll reconsider whether that term really needs to be on the list. Delete it if it doesn't.

SAFETY PRECAUTIONS

For many reasons (administrative, auditing, liability, etc.), it is beneficial to have all safety-related information in one place. However, precautions are most effectively implemented when listed throughout the procedure at the locations where the topic becomes relevant. For these reasons, the safety precautions section should function as a summary list of all the individual warnings throughout the procedure. The summary list and the individual warnings should be synchronized to each other with the exception that the summary entries will include extra wording to replace the context that was lost by extracting the precaution statement from its surrounding instructions.

For example, in the *Process Shutdown and Cleaning* section, you might warn the reader about a chemical hazard when cleaning up the workstation. In this summary section, that warning won't make sense unless you add wording at the beginning of the entry to recreate that context, such as:

> "During cleanup, when using solvent to wipe down the work surface ..."

Mechanical, Electrical, Chemical, Ergonomic

Use this section to create a bullet list of all the mechanical, electrical, chemical, ergonomic, and other precautions specific to performing the process.

Do not include general precautions related to the surrounding environment, unless those precautions are not stated in a general workplace safety document. If those are already stated in a general document, repeating them here duplicates the information, enabling one set of warnings to become obsolete or to contradict the other. It also creates the possibility of an employee following one and not the other, which could lead to injuries due to lack of critical information.

After the document content is written, running a text search on phrases within each summary entry (ones that correspond word-for-word with their warnings) will help ensure that no safety topic is left out.

NOTE: You can read more about searching for *Text Strings* on page *98*.

If all warning statements are formatted correctly, running a text search on each instance of the word *warning* will also help ensure that the summary list is complete.

NOTE: You can read more about formatting *Notes, Cautions, and Warnings* on page *84*.

Personal Protective Equipment (PPE)

Use this section to create a bullet list of all the personal protective equipment referenced in this procedure, the part number (your company's or the manufacturer's, if applicable), as well as a definition of what each item is used for.

1. Format the items in some manner that will set them apart from regular text. Consider using the following format, replacing the bracketed content with your own:

   • [*Item name*] [(Part number)]: [Purpose of the item]

2. Set up the item name (whatever comes before the parentheses in the example above) as a bookmark.
3. Insert a cross-reference to this bookmark anywhere this item is mentioned throughout the document.

   NOTE: You can read more about *Setting up Bookmarks* on page *95* and *Inserting Cross-References* on page *95*.

After the content of your document is written, come back to this section and run a text search on each item. If there is only one instance of an item referenced (the one in this section), you'll know you forgot to mention some critical information. Or perhaps you'll reconsider whether that item really needs to be on the list. Delete it if it doesn't.

SPECIAL HANDLING

Special handling involves precautions where the equipment or product is at risk of damage. Similar to the *Safety Precautions* section, the same reasons exist for having all special handling information summarized in one place as well as being listed throughout the document at the locations where the topic is relevant. The summary and the individual cautions should be synchronized to each other. Remember, the wording of the summary entries cannot be made into bookmarks, because they will vary slightly from the corresponding cautions, due to the lack of context from the surrounding instructions. Make sure to recreate that context by adding wording at the beginning of each entry, such as:

"When transporting product to the staging area ..."

Mechanical, Electrical, Chemical

Use this section to create a bullet list of all the mechanical, electrical, chemical, and other special handling requirements specific to running the process. Concentrate on instances where the equipment or product could be damaged by mishandling, but only include risks that are specific to this process.

After the document content is written, running a text search on phrases within each summary entry (ones that correspond word-for-word with their cautions) will help ensure that no handling topic is left out.

If all caution statements are formatted correctly, running a text search on each instance of the word *caution* will also help ensure that the summary list is complete.

> NOTE: You can read more about formatting *Notes, Cautions, and Warnings* on page *84*.

Quality Levels

In addition to precautions, use this section to define special requirements for handling various levels or classifications of product that are not specific to a customer. These classes should align with the *Quality Levels* defined in the *INSPECTION CRITERIA SPECIFICATION*. But instead of criteria, this section defines processing rules between classes, which may include differences in:

- Lot sizes
- Materials
- Personnel
- The quantity and configuration of setup products
- Data collection requirements
- Inspection frequency

- The use of backup equipment

EQUIPMENT

Primary and Backup Equipment

Use this section to create a bullet list of the primary equipment used to perform this process (and any backup equipment, if applicable). As a sub-bullet under each piece of equipment, insert a cross-reference to the equipment operations procedure if the machine has one. This provides a trail of reference information for anyone running or auditing the process.

Consider using the following format, replacing the bracketed content with your own:

- [Manufacturer] [Make and model] [Function of the machine]
  - o [*Document number*], [*Document title*]

Equipment is not the same as tools (which are defined in the following *Materials and Tools* section). Think of equipment as installed—or non-portable—machinery, including air hoses, fluid dispensers, and any other items installed at or near workstations where this process is performed. If this procedure references some piece of equipment, that piece of equipment should be on this list.

If there are restrictions about when backup equipment can be used, consider putting those pieces of equipment under a separate heading and defining the restrictions on their use.

If an outside vendor is qualified to perform the same task (either as a primary or backup), list the procurement specification (or whatever document governs the outsourcing) in the *Documents and Forms* section, then insert a cross-reference to that document here.

MATERIALS AND TOOLS

Materials

Use this section to list the materials added to (or installed on) the product by this process. This would be anything included on the product's BOM (bill of materials). In addition, list all the consumable items used to perform this process, such as process gasses (CDA, N_2, O_2), solvents (IPA, Acetone, Lacquer Thinner, Mineral Spirits), and materials used for cleanup (wipes, cotton swabs, etc.).

Create a bullet list of all the materials referenced in this procedure, the material's part number (your company's or the manufacturer's, if applicable), and a definition of what each item is used for.

1. Format the materials in some manner that will set them apart from regular text. Consider using the following format, replacing the bracketed content with your own:

   - *[Material name]* [(Part number)]: [Purpose of the material]

2. Set up the material name (whatever comes before the parentheses in the example above) as a bookmark.
3. Insert a cross-reference to this bookmark anywhere this material is mentioned throughout the document.

   > NOTE: You can read more about *Setting up Bookmarks* on page *95* and *Inserting Cross-References* on page *95*.

After the content of your document is written, come back to this section and run a text search on each material. If there is only one instance of a material referenced (the one in this section), you'll know you forgot to mention some critical information. Or perhaps you'll reconsider whether that material really needs to be on the list. Delete it if it doesn't.

> NOTE: You can read more about searching for *Text Strings* on page *98*.

Tools

Tools can be defined as portable—not installed—instruments used while performing the process. This may include tweezers, rubber mallets, handling aides, and a multitude of other items, depending on your manufacturing environment.

Use this section to create a bullet list of all the tools referenced in this procedure, the tool's part number (your company's or the manufacturer's, if applicable), as well as a definition of what each tool is used for.

1. Format the tools in some manner that will set them apart from regular text. Consider using the following format, replacing the bracketed content with your own:

   - [*Tool name*] [(Part number)]: [Purpose of the tool]

2. Set up the tool name (whatever comes before the parenthesis in the example above) as a bookmark.
3. Insert a cross-reference to this bookmark anywhere this tool is mentioned throughout the document.

After the content of your document is written, come back to this section and run a text search on each tool. If there is only one instance of a tool referenced (the one in this section), you'll know you forgot to mention some critical information. Or perhaps you'll reconsider whether that tool really needs to be on the list. Delete it if it doesn't.

PROCESS SETUP

We come now to the first of several instructional sections (*Process Setup, Process Operation, Process Inspection*, and *Process Shutdown and Cleaning*). Instead of lists and general information, these will present step-by-step directions for the reader to follow during setup, operation, inspection, and shutdown of the process. Because sequences are implied by these tasks, numbered lists will make up the majority of the content in these sections.

Outlining

Before writing any instructions, it's a good practice to plan your content using an outline. Start by brainstorming the high-level actions required to set up the process, such as:

- Verifying completion of the previous process
- Preparing workstations to start the current process
- Gathering tools and materials
- Turning on the equipment
- Running setup product
- Inspecting the setup product
- Etc.

For each of these actions, consider what is required to complete it. For example, *verifying completion of the previous process* might involve:

- Checking the electronic (MES system) or printed (traveler/router) manufacturing lot information to ensure that it was filled out correctly
- Taking note of any quantity adjustments from the previous process and verifying that with a physical count of the current lot quantity
- Taking note of any special instructions resulting from issues encountered at previous processes
- Discussing those issues with the personnel who performed the processes

Do this for the entire setup process. Make sure to specify when and how readers are to use other documents, such as performing an action in an *EQUIPMENT OPERATIONS PROCEDURE*, inspecting a product feature per an *INSPECTION CRITERIA SPECIFICATION*, or filling out a form. Then organize all the information into topics, subtopics, and instructions.

Flowcharting

Presenting this information to a reader in a text-based format requires organizing it into a sequential flow. This type of presentation works well if the information itself is sequential, or linear, in structure. But what if it isn't? What if your instructions have decision loops and optional content, based on the whether the product meets or fails to meet certain conditions? This is where the efficacy of a text-based presentation breaks down.

If the content cannot be easily represented by a standard outline of topics, subtopics, and indented instructions, you may need a flowchart to visually introduce your non-linear information.

If you decide to use a flowchart:

- Follow standard flowchart conventions regarding shapes and their meanings.
- Use the basic *Action Phrases* (discussed in the next subsection) as the content of your flowchart shapes, and remove all articles (a, an, the) to make the phrases as concise as possible.
- Label each flowchart shape with its corresponding section number, so the flowchart functions as a visual table of contents.
- Place the flowchart at the beginning of a section, as a way of visually introducing the content to the reader and warning them of its non-linear structure.
- The flowchart should flow left-to-right and top-to-bottom, like the text in a document.
- To ensure readability of the flowchart, limit the number of flowchart shapes across (left-to-right). If needed, the flowchart can be many rows long (top-to-bottom) to accommodate the length of the procedure.
- Present the flowchart as you would any other figure in your document.

> NOTE: You can read more about using *Figures* on page *84*.

Action Phrases

Whether or not you use a flowchart, your outline of the *Process Setup* activities will need to be refined for clarity and precision. Starting with the high-level actions (topics), edit them as follows:

- Rewrite them as concise verb phrases, instead of topics or titles (i.e. "Turn on the equipment" instead of "Turning on the equipment" or "How to turn on the equipment"). These phrases will become the headings of your instruction list. If you used a flowchart at the beginning of the section, these phrases should match those in the flowchart shapes with the addition of articles (a, an, the).

- Number each phrase to convey the correct sequence of operations to the reader. If you used a flowchart at the beginning of the section, ensure that the section numbers match the flowchart shape numbers, including any decision points, conditional content, or iterations.

- Format the phrases as second-level headings so the table of contents (which displays all first- and second-level content) will contain these abbreviated instructions, thereby functioning as a mini-procedure.

> NOTE: You can read more about *Table of Contents* on page *82* and *Numbered Text and Levels* on page *83*.

Step-by-Step Instructions

Indented beneath each verb phrase, list the instructions for accomplishing that action and edit them as follows:

- Rewrite them using active (instead of passive) language. For example, use "Rotate the power switch to the ON position," instead of "The power switch is rotated to the ON position." This may seem like common sense, but passive language often shows up in procedural documents because it is the default, conversational tone that authors use in their everyday life. In such a setting, passive language comes across as polite; in a procedure, it implies latitude where none may exist.

- Number the instructions to convey the correct sequence to the reader.

- Format the instructions as third-level text.

Other Considerations

In terms of the structure and content of the information, there are a few other considerations.

If the instructions to accomplish an action are used more than once in the sequence, don't repeat them. Define the instructions where they first apply, then insert a cross-reference to that action at the other locations in the sequence where it also applies.

NOTE: You can read more about *Inserting Cross-References* on page *95*.

If conditional instructions are so lengthy that readers will have forgotten the standard instructions by the time they get back to them, consider moving the content to an appendix at the end of the document and inserting a cross-reference to it. If the conditional instructions result from a problem (a deviation from the normal, expected sequence), move the content to the *Troubleshooting* section (discussed on page *37*), and define all the steps necessary to resolve the issue.

When pointing readers to a specific section of another document (such as performing an action in an *EQUIPMENT OPERATIONS PROCEDURE*, inspecting a product feature per an *INSPECTION CRITERIA SPECIFICATION*, or filling out a form field), use the exact wording of the section header, action phrase, criterion, or form field in that document. If such references are used often, consider creating a list of these references in the *Appendices and Miscellaneous* section (discussed on page *38*), setting them up as bookmarks, and inserting cross-references to them. Doing this maintains a one-to-one relationship between the instructions listed here and the wording used in other documents.

NOTE: You can read more about *Setting up Bookmarks* on page *95*.

PROCESS OPERATION

This section of your process procedure should be constructed in the same the manner as the *Process Setup* section, utilizing flowcharts, action phrases, and step-by-step instructions. Whereas the previous section defined how to set up the process, use this section to define how to perform the process in order to produce product.

Though similar in structure to the *Process Setup* section, this section may be larger and more complex, depending on how many machines are used, how extensive the inspections are, and the number of options or conditional paths the product may take.

PROCESS INSPECTION

The *Process Inspection* section is unique to the *PROCESS PROCEDURE*. Use it to define the evaluation activities that must be performed in association with running this process, such as:

- **Setup Inspection** – evaluation performed during setup, or on setup product, to determine if the process will produce product of acceptable quality. In some cases, the setup inspection requirements will be stricter than in-line or post-processing inspections, to ensure that product will not fail subsequent inspections if the feature in question experiences variation.
- **In-line Inspection** – periodic evaluation performed during (in the middle of) production, to ensure that product quality continues to be acceptable
- **Post-Processing Inspection** – evaluations performed after production, to verify that the manufacturing lot is of acceptable quality
- **QA Inspection** – independent evaluations performed on product to confirm that it is of acceptable quality

Create a section for each applicable activity, and do the following:

- Use descriptive titles instead of action statements for each activity (such as "In-Line Inspection" or "Width Measurement"). These are the topics that will be referenced from the *Process Setup* and *Process Operation* sections.
- Number each title to convey the correct sequence of inspections to the reader, if applicable.
- Format the titles as second-level headings so the table of contents will display them.

> NOTE: You can read more about *Numbered Text and Levels* on page *83* and *Table of Contents* on page *82*.

In addition, specify the following:

- **Inspection Tool** – microscope, calipers, etc.
- **Inspection Method** – visual inspection, measurement, go/no-go gauge, etc.

- **Inspection Frequency** – 100% or sample inspection

If the tool, method, and frequency are consistent across all inspection activities, specify them at the beginning of this section. Otherwise, specify them within each activity, where variations only apply to that specific evaluation.

Also, list the inspection criteria:

- **Process-Specific Criteria** – criteria related to a product feature that are only evaluated at this process and therefore defined in this *PROCESS PROCEDURE*
- **Process-Independent Criteria** – criteria related to a product feature that are also evaluated at other processes, or in other departments, and therefore defined in an *INSPECTION CRITERIA SPECIFICATION*. If used at this process, the document should be included in the *Documents and Forms* section of this procedure, and a cross-reference to that document should be listed here.

Finally, define what to do with defective product:

- How to dispose of defective product that **cannot** be reworked – depending on the disposal method, the instructions could be defined here, or the reader could be referred to the *Process Shutdown and Cleaning* section with the instructions defined there.
- Where to find reworking instructions for defective product that **can** be reworked – refer readers to the *Troubleshooting and Reworking* section.

If any of these inspection activities are simple enough to be defined in the *Process Setup* and/or *Process Operation* sections without interrupting the process and confusing the reader, include them in the sequence where applicable. Otherwise, insert a cross-reference to this section and define the activities here.

PROCESS SHUTDOWN AND CLEANING

This section is used to describe the activities for shutting down the process and cleaning up the workstations and materials involved.

Construct it in the same manner as the *Process Setup* and *Process Operation* sections, utilizing an outline, flowcharts, action phrases, and step-by-step instructions as applicable.

You may wish to separate the shutdown and cleaning activities or combine them into a single section. Whichever you choose, follow the same rules for wording, formatting, and numbering so the action statements are displayed in the table of contents.

Keep in mind that one of the reasons for having a separate *EQUIPMENT OPERATIONS PROCEDURE* is that equipment information is not necessarily dependent on any particular process—machines may be used in several processes. And since the equipment shutdown instructions are already defined, there is no need to repeat them here. A simple statement like the following will do:

> "Shut down the [function of the machine] equipment per the appropriate procedure listed in the *Equipment* section."

When readers turn to the *Equipment* section, they should find a list of the equipment used in this process, along with the function of each machine and the documents where they can find more detailed instructions about shutting down those pieces of equipment.

What would be appropriate to define in this procedure is the shutdown sequence for multiple machines, if more than one is involved in this process and they must be shut down in a particular order. However, if there are multiple machines but no particular order to the shutdown sequence, consider using this wording:

> "Follow the instructions in the appropriate procedures listed in the *Equipment* section to shut down the equipment."

OUT-OF-CONTROL ACTION PLAN

The Out-Of-Control Action Plan (OCAP) is another section unique to the *PROCESS PROCEDURE*. Use it to define how readers should react to problems or out-of-control situations encountered during setup, operation, and shutdown of the process. Whereas the goal of previous sections has been to define how the process is supposed to operate, the goal of an OCAP is to capture all the ways in which a process may not operate correctly. Think of this section as your information help desk; it should be the first place readers look to find out how to resolve a process-related problem, and it may refer them to another section in some cases.

There are several reasons for separating this information from the rest of the other content:

- Having a separate OCAP section encourages the author to thoroughly consider all the possible problems that may be encountered.
- It encourages the author to define appropriate and complete reactions to those problems.
- Separating the potential problems from the process helps simplify and streamline the content of previous sections, keeping them focused on how the ideal, problem-free process should function.

The content in this section can be presented in a variety of formats (numbered list, bullet list, table, etc.), but it should include the following content:

- **Problem** – a definition of the symptom, issue, or defect encountered during processing
- **Section Reference** – a cross-reference to the applicable section(s) in the process where the problem is identified. This helps readers identify the correct reaction plan for their problem when referred here from another section. This is used in conjunction with cross-references from the *Process Setup*, *Process Operation*, and *Process Shutdown and Cleaning* sections, to provide two-directional links between the problem and the reaction plan.
- **Possible Cause** – a definition of the possible cause for the problem. There may be many possible causes for a given problem, in which case, each cause should have its own reaction plan.
- **Reaction Plan** – step-by-step instructions for reacting to the problem, correcting it, or for contacting the appropriate personnel to correct it. If the correction involves lengthy troubleshooting or reworking of product, refer readers to the *Troubleshooting and Reworking* section, discussed next.

- **Responsibility** – a designation of which personnel types (Operator, Technician, Engineer, Maintenance, etc.) are authorized to carry out the reaction plan

TROUBLESHOOTING AND REWORKING

All manufacturing processes will encounter problems, and some of those problems may require detailed troubleshooting to solve. Embedding lengthy troubleshooting instructions inside the *Out-Of-Control Action Plan* could present an obstacle to readers, hindering their ability to quickly locate the solution to their problem. For this reason, put lengthy troubleshooting content here in its own section.

In addition, if the processing problems defined in the OCAP are not identified and resolved quickly, defective product may be produced. If the product cannot be reworked to bring it into compliance with the criteria listed in the *Process Inspection* section, refer readers to the disposal method defined there. If the product can be reworked, define the instructions for doing so here.

Create a section for each troubleshooting and reworking activity, and do the following:

- Use descriptive titles instead of action statements for each activity (such as "Removing Reflowed Solder" or "Trimming Ground Wires"). These are the topics that will be referenced from the *Process Inspection* and *Out-Of-Control Action Plan* sections.
- Number each title to convey the correct sequence to the reader, if applicable.
- Format the titles as second-level headings so the table of contents will display them.

MANUFACTURING EXECUTION SYSTEM

A Manufacturing Execution System (MES) is a system used in manufacturing to track the movement of product through an assembly line. At your company, this may involve an electronic system of barcoded product quantities, statistical process control (SPC) charts, and just-in-time material deliveries. Or it could be as simple as filling out some paperwork. Regardless of the method, manufacturing processes typically require some level of accounting for the quantity and quality of products moving through them.

Use this section to define the instructions for meeting those requirements and moving the manufacturing lot to the next process. If applicable, utilize an outline, flowcharts, action phrases, and step-by-step instructions (as defined in the *Process Setup* section). Follow the same rules for wording, formatting, and numbering so the action statements are displayed in the table of contents.

APPENDICES AND MISCELLANEOUS

Use the end of the document for any topics not covered in the previous sections, or reference content lengthy enough to interrupt the normal sequence of the process.

As I mentioned in the *Documents and Forms* section, you may choose to only include the required reading list there. If so, use this section to list all other referenced documents, and bookmark them so they can be used as cross-references.

Another possible use for this section is to list all the section headers, action phrases, criteria, or form fields referenced in other documents. When pointing readers to a specific section of another document, it is best to use the exact wording in that document and easiest to do so when that wording is bookmarked and inserted as a cross-reference.

NOTE: You can read more about *Setting up Bookmarks* on page *95* and *Inserting Cross-References* on page *95*.

OTHER DOCUMENT TYPES

The three essential document types (*EQUIPMENT OPERATIONS PROCEDURE, INSPECTION CRITERIA SPECIFICATION,* and *PROCESS PROCEDURE*) should be sufficient to describe the majority of the general manufacturing operations in your company. These simple pieces can be connected in a variety of ways to construct even complex systems, as you can see in *Figure 5*.

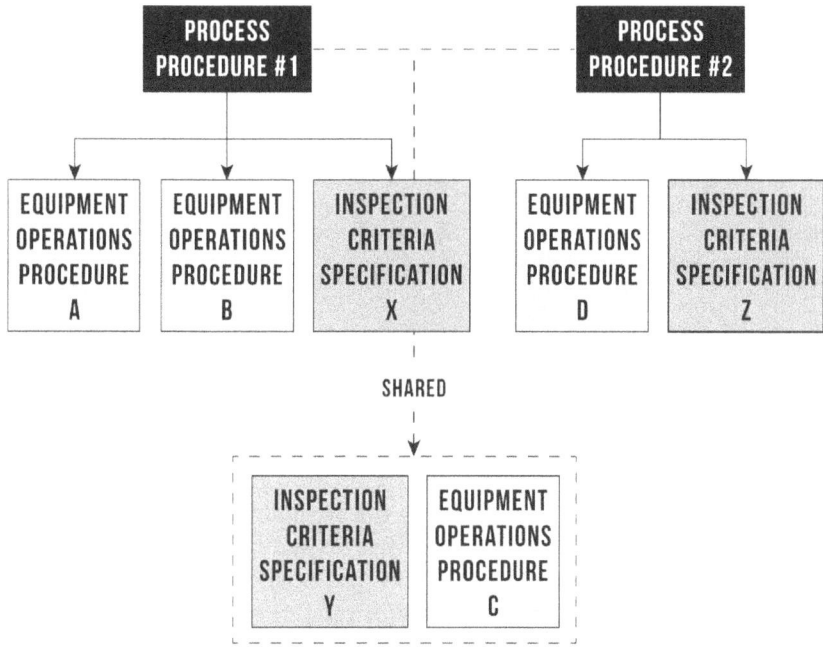

Figure 5: Building a Manufacturing Document System

With this foundation in place, we can begin to explore other document types that may be needed.

FORMS

I have not directly addressed forms until this point, because their content and format cannot be generalized. They are documents designed to capture or record information, and as such, they are specific to a particular need.

The proper recording of information requires some level of instruction. At a minimum, the individual recording the information should understand what to record, where to record it, and why it's needed. Two of the three document types already discussed are instructional documents (the *EQUIPMENT OPERATIONS PROCEDURE* and *PROCESS PROCEDURE*), making them ideal candidates for describing recording activities.

If **equipment** performance data must be collected (perhaps as part of an SPC program), and collecting this data is the responsibility of the individual operating the machine, then it makes sense that the instructions be located within the *EQUIPMENT OPERATIONS PROCEDURE*. If collecting this data is the responsibility of maintenance personnel, the instructions should be located in the maintenance documentation, which is outside the scope of this book.

NOTE: You can read more about *Equipment Programming and Maintenance Procedures* on page *77*.

If **process** performance data must be collected, and collecting this data is the responsibility of the individual performing the process, then it makes sense that the instructions are located within the *PROCESS PROCEDURE*. If collecting this data is the responsibility of QA personnel, the instructions should be located in the quality system documentation, which is also outside the scope of this book.

If **inspection** data must be collected, and collecting this data is the responsibility of the individual performing the inspection, then it makes sense that the instructions are located within the *Process Inspection* section of the *PROCESS PROCEDURE*. Again, if collecting this data is the responsibility of QA personnel, the instructions should be located in the quality system documentation.

Such data could be recorded on an **electronic** system, in which case, the procedure should direct the reader to the appropriate system.

The data could also be recorded on a **printed** form, in which case, the form could be an appendix of the same procedure that contains the instructions. Or it could be a standalone document. If you choose to make it a standalone document, consider naming and/or numbering the form as an extension (or sub-document) of the procedure. If it is an inspection form, it could also be an extension of an *INSPECTION CRITERIA SPECIFICATION*, though the instructions for its use would still be located within the *Process Inspection* section of a *PROCESS PROCEDURE*, making the paper trail harder to follow.

There may be a multitude of other forms that your company uses, but they are likely associated with business practices that are beyond the manufacturing scope of this book. The content and use of materials-related forms, for instance, would likely be governed by your company's environmental or facilities documentation. And forms related to specific customers, products, or product revisions will be discussed in the *Customer-Specific Documents* section.

HYBRID PROCEDURES

One exception to the three-document structure outlined in this book is when the process and equipment instructions cannot be separated from one another. There may be instances where an entire process consists of simply placing product on a machine and getting a result. Once the equipment operations are defined, there may be no other instructions needed to complete the process. Even if several different machines perform that same function (in completely different ways), if the general process is nothing more than pushing a button and getting a result, there may be no need to create a separate *PROCESS PROCEDURE*.

For scenarios like this, write the *EQUIPMENT OPERATIONS PROCEDURE* first. Then expand it to include whatever sections of the *PROCESS PROCEDURE* are relevant.

CUSTOMER-SPECIFIC DOCUMENTS

Once your general manufacturing practices have been documented, what remain are the exceptions. There are several options for documenting instructions specific to certain customers' products, or revisions of those products:

1. In each *EQUIPMENT OPERATIONS PROCEDURE*, include an appendix that lists any unique parameters, settings, and fixtures or tooling, organized by customer, product, and revision. Do the same for process-related information in the

PROCESS PROCEDURE and criteria-related information in the *INSPECTION CRITERIA SPECIFICATION*.

2. Create a single hybrid procedure for each product revision that lists all unique information for the entire manufacturing process, organized sequentially by process and machine.
3. Place customer-specific information on the router/traveler (the paperwork that accompanies a manufacturing lot through the assembly line).

The benefit of **Option 1** is that personnel need only search through the documentation for a single process to find the customer-specific information they need to do their job. The drawback is that all information specific to a single customer is distributed among numerous documents, which can pose challenges to personnel responsible for managing the information or those products. Although, this can be offset by clearly organizing the documentation system so one knows exactly which document should contain what type of information.

The benefit of **Option 2** is that all information specific to a single customer is located in one document, simplifying its management. The drawback is that it forces manufacturing personnel to search through all this information—most of which doesn't pertain to the process they perform—to find the information they need to do their job. This can negatively affect the efficiency and speed of the assembly line, but the drawback can be offset by clearly organizing the document sequentially by process and equipment.

The benefits of **Option 3** are that it reinforces the distinction between general and specific instructions, and it limits searching to a single section of a single document. The drawbacks are that the customer-specific information could be quite lengthy and interfere with the functionality of the router/traveler, and it would have to be printed over and over again with each manufacturing lot in a paper system.

A variation on **Options 2** and **3** would be to standardize the names of all equipment settings, inspection criteria, and process parameters. Then set up a database with this fixed structure and store the customer-specific information as variables. The information could be used to populate an abbreviated version of a hybrid procedure, or the router/traveler.

EQUIPMENT PROGRAMMING AND MAINTENANCE PROCEDURES

In the *EQUIPMENT OPERATIONS PROCEDURE* section, we discussed starting with the manufacturer's documentation and extracting the operations-related information. If you need to create a programming procedure (for engineering personnel) or a repair/maintenance procedure (for maintenance personnel), follow the same strategy for extracting and translating the information appropriate for these audiences. Then expand the document content to include any organizational knowledge gained about these topics while working with the equipment.

STYLE, FORMATTING, AND PRESENTATION

Now that we've finished discussing the structure and content of your manufacturing documentation, let's turn our attention to questions of presentation. No matter how well-written your content may be, if the presentation of it is a jumbled mess, the reader may struggle to retain the information. This is why style and formatting play critical roles in how well the content of your documentation is received.

The following sections give general, yet practical techniques for presenting your content in the clearest manner possible. For specific instructions on how some of these techniques may be accomplished using Microsoft Word, see *FORMATTING WITH MICROSOFT WORD* on page *89*.

BEST PRACTICES FOR PRESENTING INFORMATION

If followed, these best practices will greatly improve the quality of your manufacturing documentation:

- The audience should determine the type of content (format, style, complexity, etc.).
- Use scope statements at the beginning of sections to define the purpose of the section and/or to present information that applies to the whole section.
- If prerequisite information exists, present it at the beginning.
- Tell readers what they need to know; don't tell them what they don't need to know.
- Make your instructions as simple as possible.
- Structure the information logically—by chronology, by topic, or whatever organizational arrangement makes the most sense. Do not present information in a haphazard manner.
- Use active instead of passive language, unless the instruction is optional.
- Use numbered lists for sequential information—instructions that must be followed in a particular sequence.
- Use bullet lists for non-sequential information—items, or choices, that don't need to be followed in a particular sequence.

- For each instruction, separate the instruction from the result. State the instruction first, then follow that with what the result should be (discussed in detail in *Sentence Structure* on page *81*).
- Use indents to show hierarchy.
- When designing a document or a document system, first create a place for everything, then put everything in its proper place. Putting information in an illogical, non-intuitive location will only ensure that the person who needs it won't find it.
- To simplify document maintenance, pay attention to document references. In a hierarchal or parent-child document system, documents should only contain top-down references (i.e. a controlling document should reference the documents under its control). Bottom-up or lateral references may increase the number of document changes triggered by a single update. Avoid these if possible, except in cases where information must be inherited from a high-level document, such as a safety policy or other company-wide information.

FONT STYLES

Font – In typography, there are many opinions but few rules when it comes to using serif versus sans serif fonts. For technical documentation, consider the context of your organizational environment. Will your documents be viewed in print, on a screen, or both? Does your company tend to use one type of font more than another? Does the audience prefer one over another? In the absence of a precedent, try experimenting. Use one type for headers and another for body text. Style pages of content with different fonts, then place them side-by-side for comparison. Do the extra, decorative strokes of a serif font lead your eye from one word to the next and increase readability, or do they make the text look cluttered? There are many good options. So choose whatever is most readable or what is appropriate for your business, and be consistent across all your documents. However, avoid using special fonts that may not be installed on every computer, as this will cause display variations across your organization.

Size – Larger text tends to be more readable. But if your documentation will be used mostly in a printed format, large text will result in higher paper and ink usage. Whichever size font you choose, consider factors such as readability, cost, and environmental responsibility. Determine what sizes will be used for what purposes (body text versus headers, etc.), and use that consistently across all your documentation.

Color – A font color scheme can reinforce a company's identity (by matching its marketing materials) or be used to bring attention to certain portions of your documentation (instructions versus results). Again, a major consideration is whether or not your documents will be used in a printed format. If so, the same readability, cost, and environmental factors should be considered.

Bold – As you can see from this list, bold text stands out from its surroundings. Use it to emphasize content (such as warnings) or to keep certain content from being visually lost among other text (such as bullet list items).

Italics – Some style manuals recommend using italic text when referring to titles of other works, instead of placing them "inside quotes." Building on this concept, consider italicizing all references and reusable text (such as bookmarks, discussed on page *95*).

Underline – In our modern culture, a reader may interpret underlined text as a hyperlink, whether or not you intend it. Use this to your advantage. Consider underlining all references to sections of other documents.

All Caps – Beyond proper grammar, capitalization is another tool for making text stand out. Use it for headers or to keep certain content from being visually lost among other text (such as software commands).

PARAGRAPH STYLES

Left Alignment – Numbered headers and short bullet lists should be aligned to the left (not justified). This prevents the text from appearing stretched (see *Justified Alignment* on the next page) while preserving the indentation level and what that conveys to the reader (see *Indentation* below).

Center Alignment – Figures, tables, and captions are typically centered on the page, to make the best use of the horizontal space available. Center alignment is also useful for section headers, but only if those headers are not numbered.

Right Alignment – In languages that are read from left to right, this alignment type is rarely used except for header and footer information.

Justified Alignment – Justified alignment creates a clean text edge on both sides of the page, allowing the eyes to move easily from one line to the next and making the content appear well-organized. To achieve this alignment, extra space is added between words, which can make some text appear stretched. The benefits outweigh the drawbacks for most uses, so consider using justified alignment for most of your non-header, numbered text needs.

Indentation – In technical documentation, indenting is used to show information hierarchy. Indented text is subordinate to the clause above it. Or to use programming terminology, indented text is like a subroutine. A reader should be able to follow a single text level from the top of the page to the bottom, without reading the indented content, and there shouldn't be any logic gaps in the instructions. Use indenting as another way to structure your content so the reader can skip what is not relevant to their situation.

Line Spacing – Unless there is a special circumstance, consider setting the line spacing (and the amount of space before and after the paragraph) equal to the font size. For example, a font size of 12 points should have single spacing and 12 points before and after the paragraph. When used consistently across headers and body text, this strategy gives all content an appropriate amount of visual "breathing room."

SENTENCE STRUCTURE

Instructions are not the same as conversational text. In other types of writing, maintaining a reader's attention through variation is important, but in technical documentation, accuracy and consistency are the objectives. In terms of sentence structure, treat the instruction itself as primary. State it first, and place the expected result of that action (or any explanations) afterward. The only exception to this should be if the instruction must be prefaced by some other information in order to make sense.

If you wish to make the distinction between instruction and result more pronounced, consider formatting them differently from each other. For example:

1. Select the *File* tab.
2. Click the **Save** button. This will save any changes you've made to your document.
3. Click the **Print** button. This will allow you to specify a number of options for printing your document.

In this example, sentences 2 and 3 use a light version of the same font, making those comments appear less important than the instructions that precede them. The same thing could be accomplished with a color change. In a lengthy set of instructions, this type of distinction allows a reader to quickly scan content that they've already studied in detail, to refresh their memory.

TABLE OF CONTENTS

The most obvious uses for a table of contents are to inform readers about:

- What types of information are contained within the document
- Where to locate that information

There is a third use that only works if a particular strategy is followed with regard to text levels. If first-level text (with the left-most alignment) is only used for primary headers, and second-level text (with the first level of indenting) is only used for secondary headers, then the table of contents in a procedural document (that shows only these levels of text) will function as a miniature procedure.

Refreshing one's memory is how most employees will interact with manufacturing documentation on a day-to-day basis. A thorough reading is likely to occur only during initial training and certification, or after a major change, which would require retraining and recertification. For this reason, it is effective to provide the information in a summarized format.

NOTE:	You can read more about using a table of contents in Microsoft Word beginning on page *91*.

NUMBERED TEXT AND LEVELS

As I mentioned in the section on *Best Practices for Presenting Information*, numbered text is used for information that must be followed in a particular sequence, such as instructions. In addition, it's useful in a controlled document system, where changes to the content must be specifically identified for approval (i.e. "Section 4.5.2.1 was updated to reflect current manufacturing practices.").

For these reasons, and the mini-procedure concept in the previous section, consider using the following structure:

1.0 FIRST-LEVEL TEXT (for primary headers)

1.1 SECOND-LEVEL TEXT (for sub-headers)

1.1.1 Third-level text (for instructions)

1.1.1.1 Fourth-level text (for sub-instructions, explanations, and optional content)

Four levels of text should be sufficient for documenting all your headers, sub-headers, instructions, and sub-instructions. If your content does not fit into these four levels, it's probably not concise enough and may need to be rewritten and/or reorganized.

UNNUMBERED TEXT

Unnumbered text is useful for providing information at the beginning of a section, such as a scope statement that explains what is included or excluded. Because this type of information is not sequential, or instructional in nature, there is no need for numbering.

However, the appropriate level of indentation is still needed to show the reader what the information applies to. When using unnumbered text, align it to the header of the section to which it applies. If the header is numbered, align to the text, not the header number.

NOTES, CAUTIONS, AND WARNINGS

When writing manufacturing documentation, there will be instances where you want the reader to pay particular attention to something. You can accomplish this with a variety of formatting methods, but the goal is to create a visual break or obstruction that the reader cannot pass over without recognizing. One effective approach is to match the subtlety of the formatting to the importance of the content within the note. For example:

NOTE:	Use notes to bring attention to something.

CAUTION:	Use cautions to bring attention to something that could damage equipment or product.

WARNING:	Use warnings to bring attention to something that may cause personal injury or death.

Indent the note, caution, or warning so it aligns with the text to which it applies. If the text is numbered, align to the text, not the number. For cautions and warnings, also consider using symbols to identify whether the risk is mechanical, electrical, chemical, or ergonomic in nature.

FIGURES

Where possible, use figures (photos, screenshots, diagrams, etc.) to depict what is being described by the text. There are many ways to accomplish this, but the important considerations are:

- Maintaining the position of the figure in relation to the text it depicts
- Annotating the figure with text boxes, arrows, or other symbols to direct the reader's attention

- Keeping all elements of the figure together in relation to each other
- Maintaining the ability to edit the figure elements
- Naming and/or numbering the figure
- Referring to the figure

> NOTE: You can read more about *Inserting a Figure Inside a Drawing Canvas* on page *93* and *Inserting Captions* on page *94*.

REUSABLE TEXT

Some of the text in your documentation (such as document references, section references, and defined terms) will be used over and over again. You can either type each instance of these text elements manually (and commit to manually updating every instance of them if they change) or set them up as reusable text. I recommend the latter, but if you choose a manual method, use consistent wording, spelling, punctuation, and formatting for these text elements throughout your documentation.

> NOTE: You can read more about *Setting up Bookmarks* on page *95* and *Inserting Cross-References* on page *95*.

REFERENCES TO PHYSICAL AND VIRTUAL OBJECTS

Throughout your *EQUIPMENT OPERATIONS PROCEDURE* and *PROCESS PROCEDURE*, you will need to make reference to physical objects (keyboard keys, switches, and gauges) as well as virtual objects (software tabs, sections, menus, and buttons). As I mentioned in the *Table of Contents* section, it is important to finds ways of summarizing information in your documents to increase their usability. One way to accomplish this is to devise a formatting strategy for all object references that will allow them to stand out from the surrounding text. Readers wishing to refresh their memory on the sequence of buttons to push—or knobs to turn—can simply scan the text to note the order of keystrokes or objects to interact with.

The objects can be set off from surrounding text using *Font Styles* (discussed on page *79*), extra characters like (parentheses), [brackets], {curly braces}, and "quotes", or some combination of these:

- **(Power)** button
- [ON] switch
- {*File*} menu
- "***PRESSURE***" gauge

If you don't use the all-caps method, make sure to capitalize each of these references using the object's proper name (if it has one) or the name you give it using a figure annotation.

NEXT STEPS

Now that I've walked you through some high-level concepts about document system design, how documents should work together, the structure and content of the three core document types, and some formatting methods to clarify the presentation of your content, where should you go from here?

CREATE YOUR DOCUMENTS USING MY TEMPLATES

I've created Word templates for the *EQUIPMENT OPERATIONS PROCEDURE*, the *INSPECTION CRITERIA SPECIFICATION*, and the *PROCESS PROCEDURE* to be used alongside this book as a way to jump-start your progress. Each one contains:

- The sections and subsections mentioned in this book, arranged in a logical sequence and numbered accordingly
- Standardized text and writing prompts, where applicable
- Styles for numbered section headers, notes, cautions, warnings, figures, and captions
- An automatic table of contents that can be updated to reflect your content
- Headers and footers, with fields for all of your document information

These templates are preformatted with my recommended settings so you can skip the tedious work of building your documents from scratch and get right to creating the content. But you're not locked into a particular look and feel. If you want to experiment with formatting, or if your company has specific design requirements, you can modify the styles at any time to suite your needs.

These templates are available for purchase on my website, at **www.jasontesar.com**.

CREATE YOUR DOCUMENTS FROM SCRATCH

Although it would be a longer endeavor to create your documents from scratch, I wrote this book to help you do just that. Use the *GETTING STARTED* section to diagram your company's manufacturing processes and identify your documentation needs. Then use the *EQUIPMENT OPERATIONS PROCEDURE*, the *INSPECTION CRITERIA SPECIFICATION*, and *PROCESS PROCEDURE* sections to create your documentation. Consider the *OTHER DOCUMENT TYPES* that you may need. Then follow the *STYLE, FORMATTING, AND PRESENTATION* recommendations to ensure that your content is presented clearly to readers.

DOCUMENTS AS THE FOUNDATION OF YOUR TRAINING SYSTEM

Once your documentation system is established, it forms the foundation of your training system. Documents can be used for an introduction to a new process, as reference material for those in the middle of their training, and as the source material for certification and re-certification tests.

Another benefit of structuring your documents in the way I've outlined in this book is that the segmenting of process, equipment, and inspection information carries over to training in the form of flexibility and efficiency. No more training on irrelevant information just because someone put it in the same document as the relevant information. Employees only need to understand the inspection criteria, equipment operations, and process instructions for the processes they perform. And if they move to another process that uses the same equipment or inspection criteria, they only have to train on the process instructions.

FORMATTING WITH MICROSOFT WORD

The following sections provide instructions for accomplishing formatting tasks using Microsoft Word 2010. Where specific settings are mentioned, consider these my recommendations, and feel free to change them to suit your needs.

MODIFYING WORD'S DEFAULT STYLES

In the *Numbered Text and Levels* section, I discussed using the first two text levels for headings and subheadings and the next two levels for instructions and sub-instructions. Word makes this easy by supplying the styles of *Heading 1*, *Heading 2*, *Heading 3*, and *Heading 4*. Using heading styles for third- and fourth-level text (body text) may seem counterintuitive, but there are several benefits of using these four styles for all your text levels:

- The automatic *Table of Contents* (described on page *91*) is built using the text that has been formatted with the *Heading* styles.
- The *Navigation* pane automatically displays the text that has been formatted with the *Heading* styles, offering a high-level view of the document while the author is working on the details.
- Setting up a nested numbering structure between each level is easier to accomplish when all four styles are related to each other.
- These default styles can be modified at any time to suit your design needs, so they don't limit you in any way.

To change the formatting of a style:

1. Click on (select) a section of text that uses the style you want to change. On the *Home* tab, in the *Styles* section, you'll see that style outlined in yellow. You may have to scroll down to see the highlighted style.
2. Right-click on the style, and select **Modify**. A *Modify Style* dialogue box opens, allowing you to change any aspect of the formatting for that style.

CREATING A NEW STYLE

To create a new style, start with an existing style:

1. Click on (select) a section of text that is similar to what you need. On the *Home* tab, in the *Styles* section, you'll see that style outlined in yellow. You may have to scroll down to see the highlighted style.
2. Click the arrow in the bottom-right corner of the *Styles* section. A *Styles* window opens.
3. In the *Styles* window, click the **New Style** button in the bottom-left corner. A *Create New Style from Formatting* dialogue box opens.
4. In the *Create New Style from Formatting* dialogue box, change the **Name** and any other formatting.
5. Select the **OK** button. The *Create New Style from Formatting* dialogue box closes, and your new style appears in the *Styles* section of the *Home* tab.

USING UNNUMBERED TEXT

To use unnumbered text:

1. Type (or paste) your text into the document.
2. On the *Home* tab, in the *Styles* section, select the **Normal** style. The text will take on whatever formatting has been assigned to the *Normal* style (the default starting point for all text in the document, which can be modified at any time).
3. On the *Home* tab, in the *Paragraph* section, use the **Increase Indent** button to align the text with the header or paragraph above it (align to the text, not the number).

USING BULLET LISTS

To create a bullet list:

1. Type (or paste) your list into the document.
2. Select all the text in the list.
3. On the *Home* tab, in the *Styles* section, select the **Normal** style. The text will take on the formatting of the *Normal* style (the default starting point for all text in the document).

4. On the *Home* tab, in the *Paragraph* section, click the **Bullets** button. This will change the style to *List Paragraph* and insert bullets.

    a. If you wish to change the bullet style, select the drop-down arrow next to the **Bullets** button.

5. On the *Ruler*, manually adjust the **First Line Indent** and **Hanging Indent** arrows as needed, to align the bullets and text with the instructions to which they belong.

INSERTING A TABLE OF CONTENTS

To insert a table of contents:

1. Click on (select) the location where you want to place your table of contents.
2. On the *References* tab, in the *Table of Contents* section, select **Insert Table of Contents**.
3. In the *Table of Contents* dialogue box, set the *Show Levels* field to **2**.
4. Select **OK**. An automatic table of contents will be generated using all the first- and second-level header text (everything using the *Heading 1* and *Heading 2* styles).

FORMATTING A TABLE OF CONTENTS

To format a table of contents:

1. On the *Home* tab, in the *Styles* section, click the arrow in the bottom right corner to open the **Styles Manager**.
2. In the table of contents (that you inserted into your document), triple-click to select all of the first-level text on a single line. In the *Styles Manager*, the *TOC 1* style will be highlighted.
3. Right-click on the *TOC 1* style, and select the **Modify** option.
4. In the *Modify Style* dialogue box, select **Format > Paragraph**.

    a. In the *Indentation* section, set the *Left* field to **0"**, the *Special* field to **(none)**, and leave the *By* field blank.
    b. In the *Spacing* section, set the *Before* field to **6 pt.** and the *After* field to **3 pt.**

    c. Click the **OK** button.

5. In the *Modify Style* dialogue box, select **Format** > **Tabs**.

    a. If the *Left* tab is set to anything other than **0.5"**, highlight it and click the **Clear** button. Then type **0.5"** into the *Tab stop position* field.

    b. In the *Alignment* section, select the **Left** radio button. Click the **Set** button. Don't change the right tab.

    c. Click the **OK** button.

6. In the table of contents, triple-click to select all of the second-level text on a single line. In the *Styles* manager, the *TOC 2* style will be highlighted.

7. Right-click on the *TOC 2* style and select the **Modify** option.

8. In the *Modify Style* dialogue box, select **Format** > **Paragraph**.

    a. In the *Indentation* section, set the **Left** field to *0.5"*, the **Special** field to *(none)*, and leave the **By** field blank.

    b. In the *Spacing* section, set the **Before** field to 3 pt. and the **After** field to 3 pt.

    c. Click the **OK** button.

9. In the *Modify Style* dialogue box, select **Format** > **Tabs**.

    a. If the *Left* tab is set to anything other than *1.0"*, highlight it and click the **Clear** button. Then type *1.0"* into the **Tab stop position** field.

    b. In the *Alignment* section, select the **Left** radio button. Click the **Set** button. Don't change the right tab.

    c. Click the **OK** button.

UPDATING A TABLE OF CONTENTS

The content within a table of contents doesn't update in real-time, but it can be updated periodically to reflect any changes in the document.

1. Right-click on the table of contents, and select **Update Field**.

2. If an *Update Table of Contents* dialogue box appears, select the option to **Update Entire Table**.

INSERTING A FIGURE INSIDE A DRAWING CANVAS

In the *Figures* section on page *84*, I listed several topics that should be considered when using figures. The instructions in this section detail my recommended method for accomplishing this in Microsoft Word. It begins with creating a new *Figure* style, because the position (center, left, right) and paragraph spacing (before, after) of the drawing canvas is controlled by the text style where it is inserted.

1. Follow the instructions in *Creating a New Style* on page *90* to create a new *Figure* style based on the default *Caption* style.

    a. Adjust the paragraph spacing in the *Figure* style to **12 pt.** before and **6 pt.** after, and check the box for *Keep With Next* pagination.

    b. Modify the paragraph spacing in the *Caption* style to **6 pt.** before and **12 pt.** after.

    > NOTE: In a scenario where the caption is displayed below the figure, this will ensure that the figure and caption stay close to each other, do not become separated across page breaks, and have sufficient space from the surrounding text. If you wish to display the caption above the figure, reverse these settings.

2. Place the cursor where the figure will be inserted.

3. On the *Home* tab, in the *Styles* section, select your new *Figure* style.

4. On the *Insert* tab, in the *Illustrations* section, select **Shapes > New Drawing Canvas**.

5. Move the cursor to the corner of the drawing canvas boundary, and click and drag to enlarge it as needed.

    a. If the canvas is smaller than the figure being put into it, the figure will be distorted to fit. It's best to enlarge the canvas before working inside of it.

6. On the *Insert* tab, in the *Illustrations* section, select **Insert > Picture**.

    a. Or **Copy** and **Paste** the figure into the canvas.

7. On the *Insert* tab, in the *Illustrations* section, select **Shapes** to add arrows, circles, and text boxes to the figure. Sometimes a

figure is not self-explanatory, and the reader will need annotations that locate and label certain features.

8. When you're done annotating the figure, move the cursor to each boundary (top, bottom, left, and right) of the drawing canvas, and click and drag to pull them inward against the figure. The drawing canvas should not take up extra room around the figure. Because the *Figure* style was used, the drawing canvas will automatically center itself on the page.

INSERTING CAPTIONS

Each figure (photo, diagram, or table) should be labeled with a caption so that it can be referred to within the text. To insert a caption:

1. Right-click on the figure (or the border of the drawing canvas, if used), and select **Insert Caption**.
2. In the *Caption* dialogue box, enter the name of the figure in the *Caption* field.
3. Choose the appropriate *Label* from the drop-down menu:

    - **Figure** – for photos and diagrams
    - **Table** – for tables

4. Choose the appropriate *Position* from the drop-down menu:

    - **Above selected item** – to display the caption above the figure
    - **Below selected item** – to display the caption below the figure

5. Select the **OK** button.

By default, a caption inserted with the method described above will automatically have the *Caption* style applied to it. In addition, it will be numbered automatically (based on how many captions exist before it). However, the caption numbers after it will need to be updated. This is done in the same manner as updating cross-references.

NOTE: You can read more about *Updating Fields* on page *97*. Also, when referring to a figure within the document text, insert a cross-reference. Do not manually type the reference. You can read more about *Inserting Cross-References* on page *95*.

SETTING UP BOOKMARKS

A bookmark is a way of identifying a text string, section, or location in the document with a name. That name can then be used in a number of ways. For manufacturing documentation, the most useful application for a bookmark is to capture the proper wording, capitalization, spelling, and formatting of a text element so it can be used over and over again without having to re-type it (or remember exactly how it was worded, spelled, etc.). Industry terms, document references, and fields in other documents are three examples of such information.

To create a bookmark:

1. Select only the text that should be included in the bookmark. If spaces before, spaces after, and returns are not intended to be part of the bookmark, do not select them.
2. On the *Insert* tab, in the *Links* section, select **Bookmark**.
3. In the *Bookmark* dialogue box, type the bookmark name in the *Bookmark name* field.

   - Bookmark names cannot contain any spaces or special characters.
   - Because bookmarks are used to identify a variety of things, finding them within a list is easier if their names begin with the type of bookmark.
   - For a term definition bookmark, type "Definition_" followed by the term.
   - For a document bookmark, type "Document_" followed by the document number.
   - For a form bookmark, type "Form_" followed by the form number.

4. Select the **Add** button. After the bookmark has been created, it can be inserted as a cross-reference wherever that text is needed.

INSERTING CROSS-REFERENCES

When referring to bookmarked text, figures, or tables, insert cross-references to them instead of manually typing the references. In addition to the text itself, cross-references will also carry over whatever formatting has been applied to the bookmarked text. For cross-references to numbered items, figures, and tables (which are not items you've bookmarked), you'll need to format the cross-reference after inserting it, to set it apart from the surrounding text and identify it as an internal hyperlink.

To insert a cross-reference:

1. On the *Insert* tab, in the *Links* section, select **Cross-reference**.
2. In the *Cross-reference* dialogue box, select the appropriate *Reference type*:

    -   **Numbered Item** – for referring to a specific, numbered sentence or paragraph
    -   **Bookmark** – for referring to bookmarked text (such as document numbers and terms)
    -   **Figure** – for referring to figures (photos and diagrams)
    -   **Table** – for referring to tables

3. In the *Insert reference to* field, select which feature of the bookmark you would like to display:

    -   **Paragraph number** – for the *Numbered Item Reference Type*, to display the paragraph number of a sentence or paragraph. This is useful when directing readers to a specific, numbered instruction.
    -   **Paragraph text** – for the *Numbered Item Reference Type*, to display the text of a numbered item. This is useful when directing readers to a particular section of a document.
    -   **Bookmark text** – for the *Bookmark Reference Type*, to display the text of the bookmark itself
    -   **Paragraph number** – for the *Bookmark Reference Type*, to display the paragraph number where the bookmark resides
    -   **Only label and number** – for the *Figure Reference Type*, to display the label and number of a figure, such as "Figure 1". This is the preferred method for directing readers to figures.
    -   **Only label and number** – for the *Table Reference Type*, to display the label and number of a table, such

as "Table 1". This is the preferred method for directing readers to tables.

4. In the *For which caption* field, select the numbered item, bookmark, figure, or table you wish to insert.
5. Select the **Insert** button.

By definition, cross-references are connected to the items they refer to. This has benefits and drawbacks. A benefit is that the number in a cross-reference to a figure can be automatically updated, just as the figure number itself can be automatically updated. A drawback is that the cross-reference will stop working properly if the item it refers to is deleted (or sometimes, even if it is modified). For more information on both these scenarios, see *Updating Fields* (the next section).

UPDATING FIELDS

To update a single field (a cross-reference, a figure number, or a table number):

1. Select the text of the cross-reference.
2. Right-click, and select **Update Field**.
   a. If "**Error! Reference source not found**" is displayed, the cross-reference couldn't be updated due to some critical error (i.e. the referenced item was deleted or modified). Delete the bold text and any spaces on either side of it. Then re-insert the cross-reference.

To update all fields in the document:

1. Press **CTRL** + **A** to select all the text in the document.
2. Locate one of the cross-references (highlighted in dark gray). Right-click on it, and select **Update Field**.
   a. If an *Update Table of Contents* dialogue box opens, select the option to **Update Entire Table**. The table of contents is a field like the cross-references, so updating all fields in the document will affect it.
3. After all cross-references have been updated, press **CTRL** + **F** to open the *Navigation* window.

4. In the *Navigation* window, in the *Search Document* field, type the word "error." This will identify any cross-references that couldn't be updated due to some critical error (i.e. the referenced item was deleted, etc.).
5. Use the **Next Heading** (down arrow) button to jump to each location where "**Error! Reference source not found**" is displayed.
6. Delete the bold text and any spaces on either side of it. Then re-insert the cross-reference.

INSERTING FLOW CHARTS

To insert a flowchart object from another program (such as Microsoft Visio):

1. On the *Insert* tab, in the *Text* section, select **Object**.
2. In the *Object* dialogue box, under the *Create New* tab, select the appropriate **Object type**.
3. Select the **OK** button.
4. The remaining steps depend upon which type of object you're inserting and which program was used to create it. Navigate to the flowchart you created, and select it.
5. Select the **Open** button.

SEARCH-BASED EDITING

Searching is useful for identifying every instance of a text string within a document. Beyond editing for spelling, grammar, punctuation, and accuracy of the content, it is also useful for finding some common formatting errors.

Text Strings

1. Press **CTRL** + **F** to open the *Navigation* window.
2. In the *Navigation* window, in the *Search Document* field, type the word or phrase you want to find. This will identify any sections of the document that contain your word or phrase.
3. Use the **Next Heading** (down arrow) button to jump to each location where this occurs.
4. Update the text as needed.

Multiple Paragraph Returns

The correct method for adding space between paragraphs is to use the paragraph spacing settings in the *Paragraph* window. If extra returns have been used to accomplish this, do the following:

1. Press **CTRL** + **F** to open the *Navigation* window.
2. In the *Navigation* window, in the *Search Document* field, type **^p^p**. This will identify any sections of the document that contain two back-to-back paragraph returns.
3. Use the **Next Heading** (down arrow) button to jump to each location where this occurs.
4. Delete the extra paragraph returns.

Spaces at the Start of a Paragraph

In technical documentation, indenting is not used at the beginning of paragraphs as with some other types of documentation. Spaces often show up at the start of paragraphs because the content was part of another section, and when a return was added to create separation, the extra space was never removed. An editor with a keen eye may spot the misalignment this causes between paragraphs, but a more thorough method is to search for these instances:

1. Place your cursor at the beginning of the document.
2. Press **CTRL** + **F** to open the *Navigation* window.
3. In the *Navigation* window, in the *Search Document* field, type **^p** followed by a **space**. This will identify any sections of the document where a space follows a paragraph return.
4. Use the **Next Heading** (down arrow) button to jump to each location where this occurs.
5. Delete all the spaces at the beginning of paragraphs.

Too Many Spaces after a Period

Using two spaces after a period is a carryover from the pre-computer, typewriter era. Most modern style guides—including *The Chicago Manual of Style*, the *US Government Printing Office Style Manual*, and the *AP Stylebook*—recommend using only one space.

If there are some sections of the document where two spaces are intended, remove one at a time using this method:

1. Place your cursor at the beginning of the document.
2. Press **CTRL** + **F** to open the *Navigation* window.
3. In the *Navigation* window, in the *Search Document* field, type two spaces. This will identify any sections of the document where two spaces are located back-to-back. The number of matches, and a preview of each instance of a double space, will be listed in the navigation pane. In addition, the first instance of a double space will be highlighted on the document pane.

   a. If the text field does not contain the words *Search Document*, an **X** will be displayed on the right-hand side. This means that the text field already contains search text, and those search results are being displayed. Click the **X** to delete the text from the text field. Then type in two spaces.

4. Place your cursor in the highlighted area, and delete the extra space(s).
5. Use the **Next Heading** (down arrow) button to jump to the next location where this occurs.
6. Repeat this process until all extra spaces have been removed.

If only single spaces are intended throughout the document, delete all extra spaces using the following method:

1. Place your cursor at the beginning of the document.
2. Press **CTRL** + **F** to open the *Navigation* window.
3. At the top of the *Navigation* window is a text field with the words *Search Document* inside it. On the right-hand side, you'll find a magnifying glass symbol and a down-arrow. Click the down-arrow and choose the **Advanced Find** option. This will open the *Find and Replace* dialogue box.

   a. If the text field does not contain the words *Search Document*, an **X** will be displayed instead of a magnifying glass symbol. This means that the text field already contains search text, and the search results are being displayed. Click the **X** to delete the text from the text field. Then click the down arrow.

4. In the *Find and Replace* dialogue box, select the **Replace** tab.
5. In the *Find what:* field, type two spaces.
6. In the *Replace with:* field, type one space.
7. Click the **Replace All** button. This will replace all double spaces with single spaces. A window will then be displayed with the following message: "Word has completed its search of the document and has made # replacements."
8. Acknowledge the message by selecting **OK**.

    a. If there were any locations with more than two spaces, another replacement may be needed. Click the **Replace All** button again, and acknowledge the results message until the replacements number is **0**.

9. In the *Find and Replace* dialogue box, select the *Close* button.

NEED HELP?

EMAIL ME WITH YOUR QUESTIONS

Whether you start with my templates or decide to create your own, please email me (authorjasontesar@gmail.com) if you have simple questions regarding any of the topics covered in this book. I'd be happy to answer your questions if I can.

CONTRACT MY SERVICES

If you have more complex questions, want to discuss detailed implementation strategies, or need help setting up your documentation system, I offer these services on a contract basis. Please email me at the same address for rates and scheduling.

www.ingramcontent.com/pod-product-compliance
Lightning Source LLC
Chambersburg PA
CBHW061442180526
45170CB00004B/1525